Standard Method of Specifying for Minor Works

The preparation of documentation for works of repair, improvement and conversion

Third edition

Leslie Gardiner

E & FN SPON
An Imprint of Chapman & Hall
London · New York · Tokyo · Melbourne · Madras

UK	Chapman & Hall, 2–6 Boundary Row, London SE1 8HN
USA	Van Nostrand Reinhold, 115 5th Avenue, New York NY10003
JAPAN	Chapman & Hall Japan, Thomson Publishing Japan, Hirakawacho Nemoto Building, 7F, 1–7–11 Hirakawa-cho, Chiyodaku, Tokyo 102
AUSTRALIA	Chapman & Hall Australia, Thomas Nelson Australia, 480 La Trobe Street, PO Box 4725, Melbourne 3000
INDIA	Chapman & Hall India, R. Seshadri, 32 Second Main Road, CIT East, Madras 600 035

First edition 1983, Second edition 1986
published by Lewis Brooks Ltd.
Third edition 1991

Typeset in 10/11.5pt Palatino by Best-set Typesetter Ltd
Printed in Great Britain by T.J. Press (Padstow) Ltd, Padstow, Cornwall

ISBN 0 419 155201   0 442 31242 3 (USA)

British Library Cataloguing in Publication Data

Gardiner, Leslie *1917–*
Standard method of specifying for minor works. –3rd. ed.
1. Building specifications. Preparation
I. Title
692.3
ISBN 0–419–15520–1

Library of Congress Cataloging-in-Publication Data

Gardiner, Leslie, 1917–
Standard method of specifying for minor works: as in the preparation of documentation for works of repair, improvement, and conversion/ Leslie Gardiner. –3rd ed.
p. cm.
Includes index.
ISBN 0–419–15520–1 (HB). – ISBN 0–442–31242–3 (USA)
1. Dwellings – Maintenance and repair – Specifications.
2. Dwellings – Remodeling – Specifications. I. Title.
TH4817.G37 1991
692′.3 – dc20

90-42889
CIP

Contents

Section 3 Schedules of Work – Presentation and Order

Section 4 Contract

Acknowledgements

Over the several years during which this book has been in preparation there has been considerable support, much helpful comment and many contributions – far too many to list by name. I should, however, like particularly to thank: colleagues on the London South Branch QS Divisional Committee of The Royal Institution of Chartered Surveyors and members of other Divisions; architects, surveyors and other professionals among my working colleagues; and, for editorial suggestions, Martin Goodway, FRICS, and authors Lesley Joyce and the late Fergus Mackenzie, MA.

I am grateful to Messrs Colin Packington and Partners, Lague Goodway Associates and Kenneth Baker Associates, Chartered Quantity Surveyors, for assistance in preparing Section 2.

Preface to Third Edition

It is necessary to preface a third edition of this book because it is evident that the objective to promote method in the practice of specifying minor building works is slowly being recognised.

Essentially, it is aimed at improving the business of communication between builder and client, especially bearing in mind that "client" may be anyone or any group of people anywhere in the country, whilst builders have the added problem of expressing what is a technical and often a complex matter in simple terms. At the same time, the scale of the building problem nationwide is vast.

Therefore, although adoption of the method was directed at the industry in the first place, it is hoped it will also find its uses by the lay public who, after all, initiate building work.

Whereas the examples given in the book demonstrate the method, the specifier is free to specify precisely the kind of work, quality and conditions needed to provide a common document for those invited to tender.

In this edition further up-dating has been carried out, mainly with reference to British Standard Specifications or Codes of Practice, Forms of Contract, and the "Guarantee" and the "Warranty" schemes of the leading building organisations. It also introduces the requirements under the new Local Government and Housing Act 1989 concerning revised standards for housing repair and improvement grants, and how authorities will be expected to ask for a minimum of two estimates in support of grant applications – demonstrating more strongly than ever the need for consistency in builders' estimates.

The size of the outstanding house repair problem quoted in 1983 was £15 billion. The figure is now quoted as £50 billion, and on housing alone. The need for ways of streamlining and simplifying procedures is absolutely urgent.

"Time and tide wait for no man!"

Leslie Gardiner
Welwyn, Herts.
1990

Preface to Second Edition

The extent to which the first edition of this book has been taken up by the building industry and the professionals who serve it has been extremely gratifying and the early need for a second edition has provided an opportunity to make a number of changes and additions.

References to British Standard Specifications and Codes of Practice have been updated and altered where necessary due to legislation but the major addition is the extension of Section 4 to cover the Intermediate Form of Contract 1984 and BEC Guarantee Scheme 1984, which have contributed usefully to contract forms particularly applicable to minor work. In addition, the opportunity has been taken to make a number of minor amendments and additions throughout the text.

Section 2 of the book, dealing specifically with standards of materials and workmanship, should prove additionally valuable in complying with the requirements of regulation 7 of the Building Act 1984.

Many years ago the Standard Method of Measurement was devised resulting in an effective means of controlling the cost of "major" building works. Through its use items of building work are clearly identifiable and categorised, making for speedy recognition, more confident pricing and price comparison – in the manner of a modern superstore. The use of computers has become an inevitable consequence.

Measurement is of less consequence in the case of "minor" works but the clear identification of items which have more of an odd-job character and their more regular placing in a specification can be equally effective. Unfortunately, a common practice has not yet materialised and even though many individual organisations have their own standard formats, they and others all vary and the differences proliferate.

Conditions, common to the running of all jobs and the quality standards to be observed, are a logical prelude to the scheduled work. These and specification examples employing the method are included in the text.

In order to meet the growing problems of building maintenance and modernisation, there is a compelling need to re-order and streamline practices: on Council housing alone the Department of the Environment's current estimate of immediate expenditure is reported to be £15 billion and, of course, maintenance is ongoing. Additionally, the work on private housing and all other types of building could well exceed that figure.

The Office of Fair Trading has called for improved practice to protect the public and the Building Employers' Confederation is currently promoting a "Guarantee Scheme" to protect its members' clients.

The work is the culmination of a lifetime's involvement in the subject and the unique opportunity to carry out studies nationwide of specification practices.

The Collins dictionary describes Method as "the technique or arrangement of work for a particular field or subject" and in specification practice, so be it!

Leslie Gardiner
Welwyn, Herts
January 1986

Preface to First Edition

The Romans had a saying, *multum in parvo*, i.e. much may be contained in a little. If I may apply this to the construction business it could be said that much hassle, and unnecessary hassle, is currently incurred through "small" or "minor" works documentation. Yet the problems it raises in respect of the repair, conversion and/or improvement of buildings is one of importance not only for the builder but for the paymaster, whether government, local authority or property owner. I have seen a vast cross-section of documentation during a lifetime in private practice and public departments, prompting me to devise a more methodical and ordered approach. I therefore suggest, in this book, ways of reducing the present haphazard, indeed chaotic, state of dealing with the matter.

The aim is to provide an ordered basis for pricing work and cost control, as well as to suggest criteria for presenting information to an owner about a project and for moderating the risks entailed in the investment.

To avoid the kind of inadequacies and misunderstandings that arise there should be:

> a sufficiency of information about matters common to works in general and to materials and workmanship that will ensure acceptable standards of quality and performance suited to the job in hand;
>
> consistency in presentation of the works and their locations through a system of established conventions relevant to specification in the way that they have become established for measurement through standard methods of measurement;
>
> the adoption of one of the recognised published Contract Forms to suit the job requirements.

There should also be less casual use of everyday terms through the adoption of a comprehensive set of definitions.

Documentation may additionally include drawings and, if a job is large enough, bills of quantities; but for all works, particularly those involving alterations and repairs, the written word is necessary. Accordingly, the specification prevails as an essential document capable of fully informing the different parties – whether concerned with the job's finance or achievement.

Many organisations have found it necessary to produce standard specifications but unfortunately these all vary. This book does not attempt yet another – instead it aims to promote METHOD, offering a standard method and specification akin to the standard method of measurement for preparing bills of quantities. No doubt many practitioners will be able to improve on descriptions offered in this book but it is hoped that this is not a matter that will obscure the objective of promoting method in specifying.

"Let there be light" said the Good Book – which, if not referring specifically to the arcana of the building industry, does provide a cautionary note not altogether inept in the context.

"Let all things be done decently and in order."

Leslie Gardiner

Foreword

I was glad to be able to write forewords both to the first edition of Leslie Gardiner's valuable book in May 1982 and to the second edition in December 1985. It is a testimony to the value of his work to the building industry that his book has now reached its third edition in seven years.

I am even more delighted to support this third edition. Improvement, conversion and repair work usually involve direct contact between the client, a private householder, and the builder. There are too many "cowboys" around in this sector of the industry, as has been pointed out in several reports by the Government and the Director-General of Fair Trading. Several Trade Associations have sponsored consumer protection/guarantee schemes to help the customer, and this is a very welcome development.

The particular value of Leslie Gardiner's book is that it sets out detailed and formal specifications for this type of work. This is of great help to builders and clients, and I am very pleased to commend this third edition accordingly.

Michael Latham, M.P.
House of Commons
1989

How to use this book for specifying

The content of this book, arranged in four sections, is set out in the Contents List at the beginning of the book. The rules for specifying are mainly in Sections 2 and 3.

Section 1 provides two examples of Schedules of Works and these start with a statement as to matters commonly applying to both – i.e. they are prepared in accordance with a standard practice and with a set of clauses covering general contractual conditions, etc., referred to in the heading. Being extensively repetitive, these are equally applicable to other jobs and can be adopted by reference in a similar way. If convenient, they may be made available for inspection, for instance at the offices of the client's consultant, local authority, housing association, property manager, residents' association, etc. Where amendments to a standard version need to be made for the job in hand, these can be specified in the schedule of works.

There then follows a statement as to the parties (client, builder, etc.), addresses of property, a brief description of the project, references to drawings (if any), Form of Contract (types of which are given with their respective clause headings in Section 4) and so on.

The actual works are next to be scheduled and follow the rules contained mainly in Section 3. This provides for specifying works in a consistent order set out under the heading "Standard Order for Scheduling Work" (page 143). The sections immediately following define the make-up (i.e. the parameters) of the many separate parts of a building, to be used as standard.

Sections 2 and 3 are comprehensively compiled and designed to cover extensive alteration and repair work. In use, the listed contents of these sections additionally form checklists, as well as permitting the selection of items for smaller jobs.

Different kinds of jobs need to be done to parts of a building at different locations; they may entail complete renewal, partial renewal, repair, removal and so on. The various alternatives are also defined in Section 3.

To illustrate the way in which this works in practice, take for example the work to some windows. These are generally of wood or metal: thus, to ensure a *standard of quality*, clauses for woodwork, steelwork and/ or metalwork will apply as in Section 2. The clauses relating to the *execution of the work* will be found in Section 3. If repair is involved, a clause from Miscellaneous Sundries and Repairs (page 183) applies. The work may also call for a new opening to be formed, or for a window to be removed and the opening filled in: bricks, sand, cement, etc., may be required and specified per Section 2; their execution, per Section 3. Scaffolding, if needed, will have been covered by General Matters (page 69).

Descriptions of certain operations such as removal of rubbish, forming or filling openings, overhaul, etc., that can usefully be standardised have been included at the end of Section 2 under the heading "Usage of Sundry Building Terms". Terms used casually every day, such as "improvement", "modernisation", "restoration", etc., have been defined in a Glossary at the end of Section 3 (page 195).

A type of Master Schedule for use on several properties is given on page 64.

In the interests of brevity certain qualifications leading to the performance required have been omitted. In other words, simple statements of requirements may be given in more precise terms by the specifier. These will therefore have to be considered for the job and circumstances in question.

Section One

Introduction and Examples

SCHEDULES OF WORKS

Introduction

There is no widely recognised method for producing specifications of repair improvement and conversion work to buildings, yet they are the main document for communication between parties concerned, private and public, including public authorities when the work entails making grants, seeking subsidies under various Acts, dealing with VAT and so on.

Specifications in practice vary widely in their presentation, ranging from simple lists of items to sets of fully detailed documents. The former are inadequate for the needs of all parties; the latter contain repetition of published building regulations, standards relating to the quality of materials, etc. and other common contract matters, lengthy descriptions, schedules, numerous drawings and bills of quantities, which together contribute inordinately to costs of production.

Although buildings consist of numerous parts, there is no standard practice for dealing with such parts in any sort of order or recognised groupings by which they may be commonly identified. Often builders have to "spot price" items made up in different ways and, in the absence of any consistency, to adapt to every alternative.

Builders themselves are frequently called upon to estimate and specify work for clients, involving time-consuming "desk work". A consistent practice could assist the presentation of proposals to the client, explaining alternative solutions and dealing with the authorities.

At the same time, specifications need to be comprehensive to satisfy various interests and be legally competent, hence method is required. Meanwhile, the many inconsistencies continue to slow down the whole communication process.

A specification method is presented herein taking the above matters into account and following this Introduction two examples of how it works are given – one for the improvement of a house, the other the conversion of a building into two self-contained dwellings. The examples are given early in the text to show the directness, simplicity and similarity in the arrangement of two specifications showing the ease with which to find one's way about them although the projects are different. This is because of the order ensuing from the basic concept that all buildings have the same readily identifiable parts, or elements, namely walls, floors, roofs, rooms, common parts and services which the method simply extends by defining the composition of the parts (termed, for the purpose, 'parameters') making up a building. Any or all are subject to repair, alteration and/or finish which along with relevant matters are ordered in a methodical way.

The method is fully described in Sections 2, 3 and 4 and ends with a Glossary of Terms giving definitions of terms otherwise used casually everyday.

Section 2 covers matters common to most building contracts – the Preliminaries and Preambles. These include descriptions of materials and workmanship covering all trades likely to be employed. If they should prove more comprehensive than a job requires, no extravagance is incurred; if not sufficient in scope, any amendments or additions would necessarily be made. It is not unusual for a Section such as this to be compiled comprehensively and used repeatedly duly amended for the job-in-hand. Organisations may thus compile it to meet the quality standards they require and save repetition on successive jobs by simply referring to it, issuing amendments if need be to suit a particular job.

Section 3 deals with the works themselves and adopts a logical order for both locating and describing them. Starting with those to make the structure intact – working from the outside to the inside of a building – it is followed by any structural subdivisions, fitting-out, finishings and services. Because no recognised practice exists, it comprehensively classifies building parts and types of building operation that can be carried out to them. For example, a bath is deemed to comprise the bath itself, trap, waste pipe, taps (including their "tails" or service connections) and side panels, whilst operations may entail a new installation, renewal, or partial renewal. Working conditions, safety and stability, what to do about rubbish and related general matters are covered. The location of work, its nature, how it is described and general matters that surround every item of work are therefore dealt with in this Section.

Item descriptions depend on the nature of the operation. Normally, new work will need to quote size, type, quality of materials, and describe the work entailed, whereas items to be renewed, or partially so, can refer to existing parts for their description. The two examples demonstrate the method in practice, showing the brevity in item content, order, and consistency in presentation.

The order established for specification can also be followed for the building survey.

Many terms cause confusion by their casual use; for example, repair may refer to renewal, ongoing maintenance, or consequential making good occasioned by other work. Repair may also include the replacement of broken or obsolete parts, restoration due to wear and tear, incidental work to complete another piece of work and loosely, to modernisation and improvement work. It is even used to refer to the rebuilding of a major structural part. A set of definitions is therefore contained in the Glossary of Terms at the end of Section 3.

All types of formal contract and tender are used on works of repair, improvement and conversion. But since numerous jobs proceed happily from inception to completion – on a builder's quote and an implied acceptance – informality is also prevalent but is not recommended and public authorities would certainly not proceed in this way. The Contract Forms available provide a choice. They are published by the Joint Contracts Tribunal and comprise:

Standard Forms, with and without quantities, including local authority editions.

Agreement for Minor Building Works.

Agreement for Renovation Grant Works where a grant is made under the Housing Acts and an architect/supervising officer is appointed.

Agreement for Renovation Grant Works where a grant is made under the Housing Acts and where no architect/supervising officer is appointed.

Intermediate Form of Building Contract – for works of simple content.

Section 4 contains a tabulation of the clauses included in each of the above forms to assist in choosing the one most appropriate. Also included in this Section, is an account of the Building Employers Confederation "Guarantee Scheme" and Federation of Master Builders "Warranty Scheme".

Where several somewhat similar properties comprise a project, Master Schedules can be produced applying the method by using paper ruled with feint lined columns. A sample page of such a schedule is given following the Examples. The same approach can be employed to locate work where jobs are to proceed on bills of quantities prepared in trade form, or for recording quantities at various locations on survey. This approach can also assist co-owners (such as Residents Associations), landlords, estate owners, property managers, housing associations and local authorities either on their own properties or in the use of their

Agency Powers under Section 75 of the Housing Act 1969 by their undertaking to schedule the particular requirements of several owners and combining them in a contract. Similarly, schedules can be used on groups of houses when combined in a Housing Renewal Area.

In respect of drawings and structural calculation sheets, it is suggested the aim should be to limit these to A3 size or, wherever possible, A4.

Where there is disorder, quality suffers and the general aim should be to make for greater consistency. To sum up therefore, consistency in the production of documentation is a necessary objective, because the field of 'small' or 'minor' works is large and diverse leading to many different – and indifferent – practices. It impedes communications in building projects whatever their size, furthermore it detracts from their legal merit because in any form they invariably constitute a contract between parties. It also inhibits builders' pricing firmly for such work which in turn restricts the amount of useful information on costs by which to weigh the economic factors when considering investment in alternative forms of building development, particularly with the ever growing trend towards 'rehabilitation' of housing and upgrading of buildings to modern requirements in lieu of building anew.

Schedules of works

Example "A"

This scheme is a hypothetical IMPROVEMENT and REPAIR of a TWO STOREY TERRACE HOUSE.

Example "B"

This scheme is a hypothetical CONVERSION and REPAIR of a THREE STOREY TERRACE HOUSE into a FLAT and MAISONETTE.

NOTE: 1. Detailed sizes and particulars quoted are for the purposes of example only; where they are in common usage or are matters of choice (such as the type and colour of fittings, etc) these are included in brackets, otherwise they are dimensioned and quoted as may suit the examples.

2. Guidance notes to the use of SMSI in the Examples are given in italics.*

* *Standard Method of Specifying for Minor Works*

The SCHEDULE OF WORKS is prepared in accordance with the Requirements for Works of Repair, Improvement and Conversion'' dated to which reference should be made for general contractual conditions; requirements in respect of materials and workmanship and of the items of work scheduled. The Requirements can be inspected at the offices of:

Architect/Surveyor or Other Consultant.

OR

elsewhere – to be stated.

Preliminaries and General Matters

Note: Particulars for either example will need to introduce the parties to the project, address of property, brief description of works, etc as follows:

Parties (Names and Addresses):

Employer:

Architect/SO:*

Builder (where no Architect/SO):

Quantity Surveyor: (if appointed)

Address of Property:

Description of Works:

Ex. "A": The scheme is a hypothetical improvement and repair of a two storey terrace house.

(Note: The house is occupied and is expected to be so during the execution of the works for which due allowance should be made in tendering).

Ex. "B": The scheme is a hypothetical conversion and repair of a three storey terrace house into a ground floor flat and upper maisonette.

(Note: The house is vacant at present and will not be occupied until the completion of the works).

**SO = Supervising Officer*

Drawings: Ex. "A": see end of Schedule of Works for Ex. "A".

Ex. "B": see end of Schedule of Works for Ex. "B".

Form of Contract/Agreement: Ex. "A" & "B"

Enter here the title of the Form of Contract, or Agreement, to be used. *See Section 4 for information on various alternative types.*

Information required by Contract/Agreement Form Clauses: *The following are based on the Agreement for Renovation Grant Works, as an example:*

Clause 6(i)

Date for Commencement: *enter date*

Note: This date entered on invitations to tender will give the builder a firm date for pricing his tender and to assess the duration of the contract.

Date for Completion: *enter date ascertained*

Clause 6(iii)

Working Hours: *enter times*

Clause 7(ii)

Employers and Public Liability Insurance
Cover: £ *usually % of tender value*

Clause 8(i)

Fire Insurance – Existing Structure and New Additions:
At sole risk of the employer; a policy will be taken out by the employer £ *to the value of the property when completed before the works commence.*

Keeping the Works Clean: Ex."A" & "B"

Debris arising from demolition, taking down, stripping, etc is to be removed to a point for loading and for prompt disposal off-site to tip or store unless required to be retained for re-use in the work, or otherwise described to be dealt with. The builder is to take all precautions to prevent wastes, gullies, drains and manholes from receiving debris and waste materials and to leave clear and clean on completion.

Contingencies: Ex."A"

Include a provisional sum amounting to approximately 5% of the tender value for Contingencies: see summary.

Dayworks: Ex. "B"
(alternative to "A")

Include a provisional sum amounting to approximately 5% of the tender value for Daywork based on time, materials, plant, profit, etc.
see summary.

EXAMPLE A

IMPROVEMENTS

INVOLVING THE INSTALLATION OF HOUSE AND – ILLUSTRATION OF BUILDING

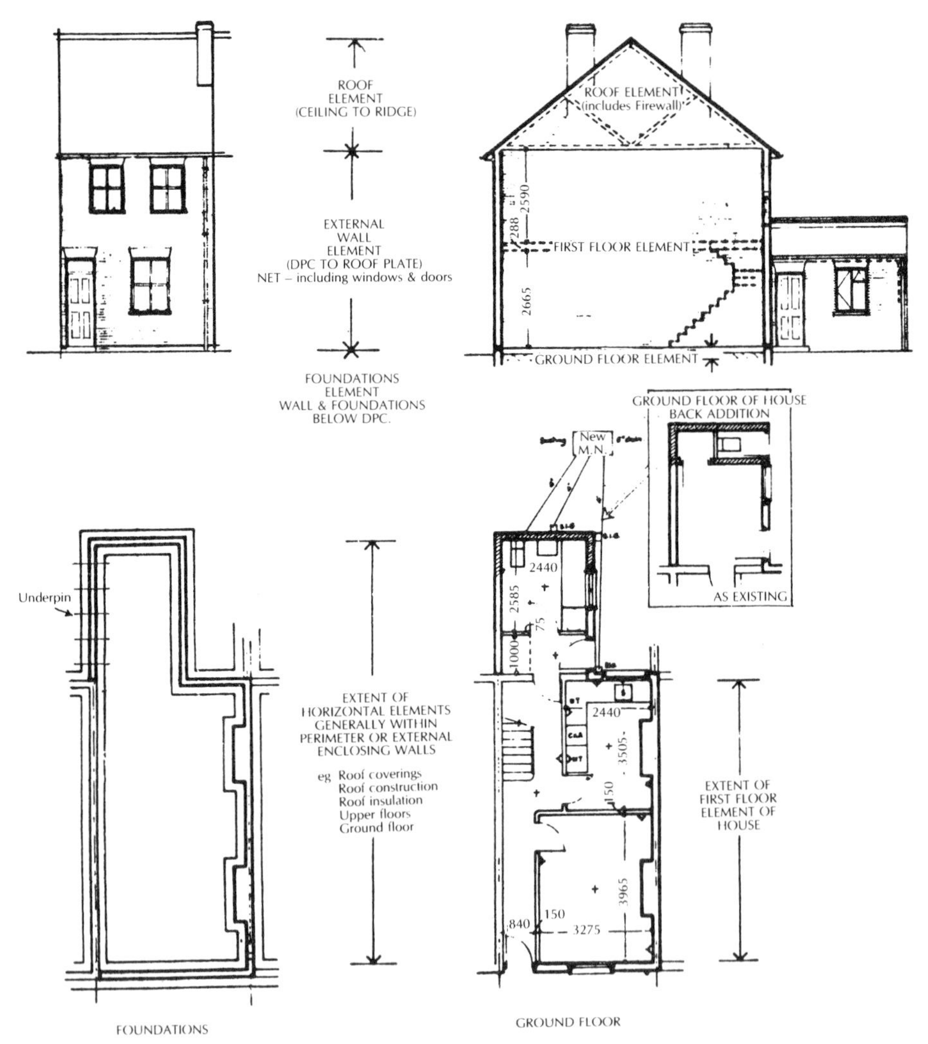

Dimensions in mm.

EXAMPLE **A**

KITCHEN AND BATHROOM TO A TWO STOREY ELEMENTS SHOWING THEIR EXTENT (OR "PARAMETERS")

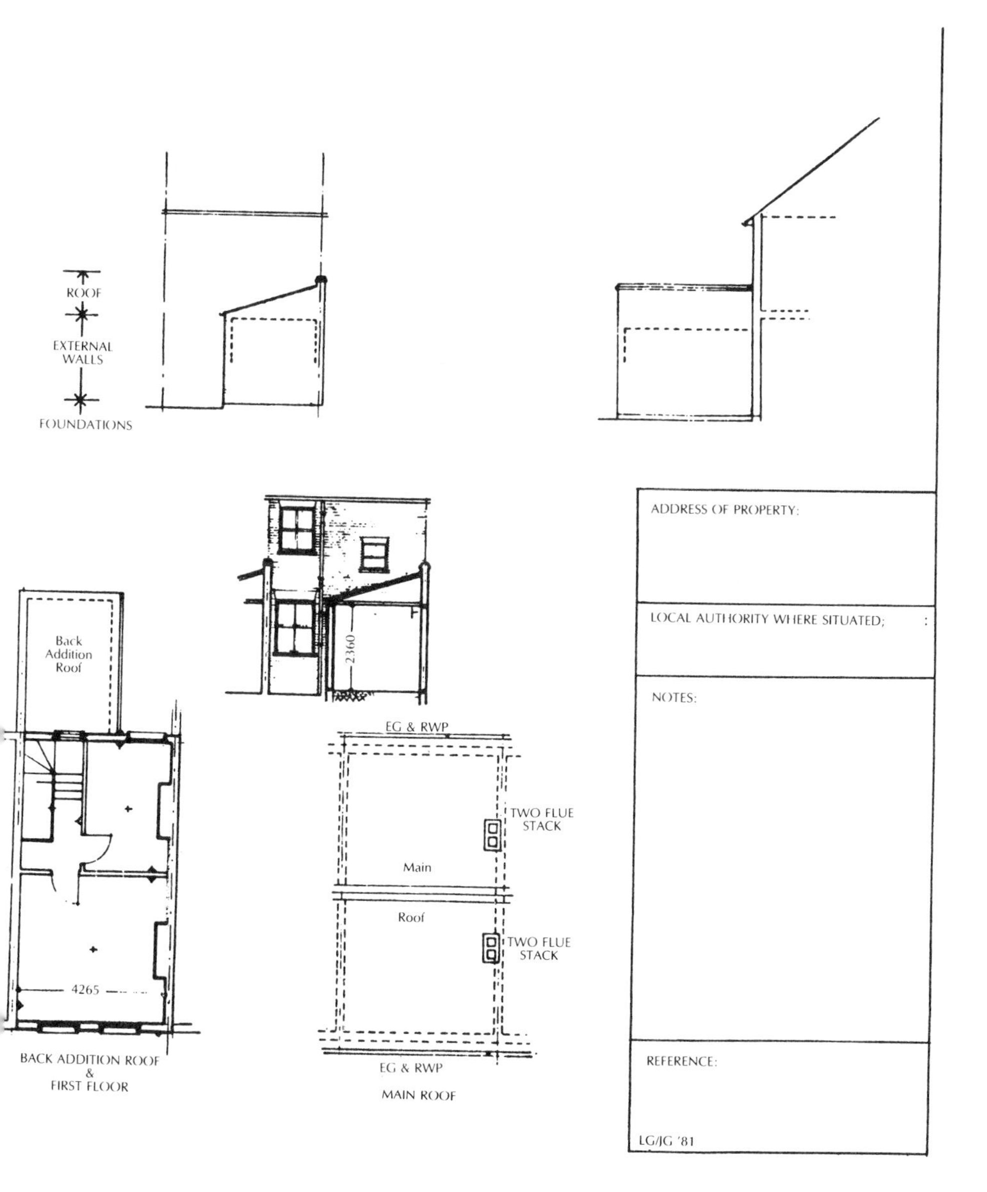

Example A – Schedule of Works

	Items £ p	Section Totals £ p

EXTERNALLY

Underpinning

A Underpin the wall on the party wall side of back addition approximately 3m long to a firm foundation – estimated at not exceeding 1m deep below existing floor level.

Demolition and Pulling Down

B Strip slates from roof to lean-to at end of back addition (containing WC and Store) demolish brick walls and grub-up foundations.

C Pull down the end brick wall of the back addition and provide temporary screen.

D Retain slates for reslating and brick rubble for hardcore.

	Items		Section Totals	
	£	p	£	p

External Walls

Isolated Openings

A Take out window in back addition and enlarge opening with jambs and head squared-off to receive 920mm wide × 1225mm high window.
Note: New jamb one side is formed in Extension.

B Form window opening with rebated jambs and head in rear brick wall of main building at height above staircase as directed on site to receive 610mm wide × 762mm high window.

For windows – see internally.

Repairs

C Cut out and stitch-in whole bricks along the two cracks in the rear brick wall of the back addition approximately 1.25m long each in matching bricks.

D Rake out and repoint all existing faces of brick walls of back addition.

E Ditto rear wall of main building.

Roofs

Roof Coverings

F Renew slated covering to slopes of main roof on new battens and felt.

	Items		Section Totals	
	£	p	£	p

Roof Coverings continued

A Ditto back addition roof including renewing soakers at abutment with rear main wall with cement fillet and tile listing. Re-edge and repoint flashing at top edge.
Note: This roof to be extended – see "Extension".

Eaves, Gutters and Downpipes

B Renew gutter to eaves at rear of main building in (100mm PVC) and downpipe in (75mm PVC).

C Renew gutter to eaves and downpipe to back addition as for main building.
Note: Gutter to be extended – see "Extension".

Repairs

D Repoint both twin flue chimney stacks, reset chimney pots and renew flaunchings.

E Overhaul gutters and downpipes to front of main building, renew defective lengths and brackets.

Insulation

F Lay over the ceiling joists in roof spaces of the main and back addition roofs (one layer of 75mm fibreglass quilt).

Extension

Extend the back addition approximately 1m from the existing rear wall by full width approximately 3m with:

		Items		Section Totals	
		£	p	£	p

Extension continued

A	Foundations: 825mm deep below existing ground level comprising concrete strip and brick footings and felt damp proof course.
B	Ground Floor: Made out in the same solid construction and finish as the new floor to the back addition – see Internally – Bathroom.
C	External Walls: In brickwork to end wall and both return walls to match existing in thickness, bond, external finish and heights and incorporating jamb to enlarged window (see External Walls – Isolated Openings) and with raked out joints Internally.
D	Roof: Made out in timber construction, covered with slates on battens and sarking felt matching the back addition and including extending parapet wall and flashings.
E	Eaves: The fascia and soffite made out to match existing and the gutter in PVC as previously described under Eaves Gutters and Downpipes.
	For plaster, window, etc – see Internally.

Sundry

F	Fit insert of appropriate diameter to 3 No. chimney pots on discontinued chimney flues.

		Items £ p	Section Totals £ p
	Paintwork		
A	Prepare and paint (one) undercoat and (one) gloss finishing coat to all existing wood and ironwork.		
B	Prepare and paint (two) undercoats and (one) gloss finishing coat to all new wood and ironwork.		

	Items		Section Totals	
	£	p	£	p

INTERNALLY

Stripping Out

A Take out all fitments, shelving, battens, skirtings and other redundant items from back addition.

Floor

B Take up flag paved floor in back addition, excavate to required depth to bring new solid floor up to general floor level of house and lay new floor to Extension consisting of 100mm hardcore, 1200g polythene membrane, 150mm concrete with trowelled finish and 2mm PVC vinyl asbestos floor tiling fixed with adhesive to manufacturers recommendations.

C Thicken out concrete floor for width of 500mm under line of new partition across back addition as foundation.

D Build-in airbrick in new end wall and lay 100mm clay drain pipe ducting from this wall under new floor to rear main wall to ventilate under floor of house.

DPC in Existing Walls

E Install damp proof course to all existing house walls and partition walls by *Method to be stated* *(eg silicone injection; insertion of slates; etc.)*

		Items	Section Totals
		£ p	£ p

Partitions

A Build stud partition across back addition between bathroom and lobby about 3m long and full storey height of 2362mm with plasterboard covering both sides and with one door opening.

B Ditto front to linen/cylinder cupboard about 1m long with ditto and ditto.
For plaster finishes, doors, etc see later.

First Floor

BR1:

C Block-in fireplace opening.

D Electrics – see schedule later.

Repairs:

E Renew plaster to ceiling.

F Overhaul double hung sash window.

G Ditto door.

H Decoration – see clauses later.

BR2:

I Block-in fireplace opening.

J Electrics – see schedule later.

Repairs:

K Renew plaster to rear wall.

L Ditto ceiling.

		Items		Section Totals	
		£	p	£	p
	Repairs continued				
A	Overhaul double hung sash window.				
B	Ditto door.				
C	Decorations – see clauses later.				
	Ground Floor				
	LR1:				
D	Take out fire surround and interior and adapt opening to suit new back boiler installation (see Heating later) and install approved fire surround and hearth PC £				
E	Take out the two cupboards in recesses each side of chimney breast and make good.				
	Repairs:				
F	Repair plaster to all four walls affected by rising damp.				
G	Renew floor boarding where defective on the fireplace and window sides of the room for approximately one third of the floor area, renew unsound supporting floor timbers and treat with wood preservative.				
H	Overhaul double hung sash window.				
I	Overhaul door.				
J	Decoration – see clauses later.				

		Items £ p	Section Totals £ p

Kitchen (LR2)

A Block-in fireplace opening.
Install the following kitchen fittings from Messrs range:

B Sink unit – 1 No. each size 1400mm long × 533mm deep × 902mm high.

C Floor units – 4 No. each size 533mm long × 533mm deep × 902mm high (or equivalent volume) with tops and backrail.

D Allow for worktops and backrail approximately 600mm long to provide continuity over space between floor units.

E Broom cupboard – 1 No. each size 533mm long × 533mm deep × 1981mm high.

F Wall cupboards – 1 No. each size 1067mm long × 321mm deep × 571mm high.
– 1 No. each size 533mm long × 321 mm deep × 571mm high.

and from Messrs range:

G Combined stainless steel sink and drainer 1400mm long with plug and chain, Hot and Cold pillar taps connected with copper tails to services, plastic overflow, deep seal trap and waste pipe jointed to inlet of gulley externally.

H Tee off above stop cock on rising main in kitchen and run connection in copper to point under new sink.

		Items		Section Totals	
		£	p	£	p

Kitchen (LR2) continued

A Radiator – see Heating schedule later.

B Electrics – see schedule later.

Repairs:

C Renew plaster to all four walls affected by rising damp.

D Renew remaining plaster on rear wall.

E Renew floor boarding all as described for LR1.

F Overhaul double hung sash window.

G Ditto door.

H Decorations – see clauses later.

Bathroom and Lobby (ex Back Addition)

I Make out plaster to new walls of ditto.

J Ditto plaster to new ceiling of ditto.

K Finish both sides of plasterboarded partitions with plaster setting coat.

L Fit two light wood window 920mm × 1225mm in prepared opening, with opening light and fanlight both hung on steel butts and fit SAA stays and SAA fastener to opening light. Fit window board.

	Items		Section Totals	
	£	p	£	p

Bathroom continued

A Fit frame for and hang 35mm hardboard faced both sides internal door 762mm × 1981mm to bathroom on steel butts and fit mortice locking latch and SAA lever handled furniture.

B Fit frame with hardwood cill for and hang 44mm two panel type 2XG inward opening external door 762mm × 1981mm to lobby hung on steel butts and fit mortice lock latch, 2 No. 150mm SAA bolts, SAA Weather bar and glaze with Group 2 obscured tempered glass with beads in putty.

C Fit frame for and hang 35mm hardboard faced internal door 686mm × 1981mm to linen cupboard on steel butts and fit bales catch and SAA knob furniture.

D Fit two slatted wood shelves full depth of linen cupboard.

E Run 12mm × 75mm wood skirting to all walls and partitions.

Install the following sanitary fittings from Messrs range in (colour):

F Bath 710mm × 1700mm on adjustable legs with approved side panels, " " Hot and Cold taps and copper tails to services, plastic overflow and deep seal trap and waste pipe jointed to inlet of gulley externally.

		Items £ p	Section Totals £ p
A	Basin 559mm × 406mm on brackets with " " Hot and Cold taps and copper tails to services, plastic overflow and deep seal trap and waste pipe jointed to inlet of gulley externally.		
B	WC suite with P trap jointed to drain, double flap plastic seat in (colour), low level " " cistern with pressure ball valve and silencer tube, plastic flush pipe in (colour) and overflow.		
C	Cylinder (in linen cupboard) – see Hot and Cold Water schedule later.		
D	Towel Rail – see Heating Schedule later.		
E	Electrics – see schedule later.		
Sundry:			
F	Cut or leave hole in new rear wall at high level fitted with airbrick externally, and vent internally.		
Repairs:			
G	Renew plaster to remaining three walls of back addition affected by rising damp.		
H	Renew remaining plaster to last.		
I	Renew plaster to existing ceiling.		
J	Decorations – see clauses later.		

		Items £ p	Section Totals £ p
Staircase, Landing and Hall			
A	Form opening in landing ceiling and fit 25mm thick hardboard faced both sides trap door 600mm × 600mm.		
B	Fit double hung sash window 610mm × 762mm with spring balances in prepared opening in rear wall generally to match type and construction of existing windows.		
C	Radiator (in Hall) – see Heating schedule later.		
D	Electrics – see schedule later.		
Repairs:			
E	Renew plaster to all walls and partitions affected by rising damp.		
F	Renew floor boarding in Hall where defective from entrance door to stairs and to cupboard under stairs, renew unsound floor timbers, treat with wood preservative.		
G	Resecure handrail to wall on landing and make good plaster.		
H	Overhaul entrance door.		
I	Ditto door from lobby.		
J	Decorations – see clauses later.		

	Items		Section Totals	
	£	p	£	p

SERVICES

Cold Water Supply

A Divert existing incoming (lead) cold water supply now terminating in back addition to a point in kitchen and install stop cock then extend as rising main in copper and connect with CWS tank in roof space.

Hot and Cold Water Installation

B Install on bearers 50 gallon CWS tank with ball valve and plastic overflow run to point externally at eaves.

C Run Cold down supply in copper to bath, basin and WC and to HW cylinder in linen cupboard.

D Install on bearers 30 gallon indirect HW cylinder with boss for immersion heater and run copper expansion pipe to discharge over CWS tank.

E Run Hot down supplies in copper to sink, bath and basin.

F Run flow and return in copper from back boiler in LR1 to HW cylinder.

		Items		Section Totals	
		£	p	£	p
Hot and Cold Water Installation continued					
A	Install insulation jackets to CWS tank and HW cylinder and insulate all pipes in roof space and under floors.				
B	Install isolating stop cock on Cold down service.				
C	Immersion heater – see Electrical schedule later.				
Heating Installation					
D	The installation is to provide space and water heating. Space heating to reach 21°C when the outside temperature is 1°C and falling.				
E	Heating is by room heater/back boiler installed in LR1 additionally serving radiators in kitchen and hall, towel rail in bathroom and HW cylinder in linen cupboard.				
F	Provide attendance on installer, arrange for connection to incoming mains and for metering. Ensure satisfactory completion in sound and working order and pay charges.				
Gas Installation					
G	Arrange with the Gas Board or their approved installer to provide points adjoining fireplace in LR1 and at cooker position in kitchen.				
H	Provide attendance, etc as described for Heating Installation above.				

	Items		Section Totals	
	£	p	£	p

Electrical Installation

A Renew the electrical installation to meet current standards and the following requirements:

	Lighting	Switch Socket Outlets	Heaters	Cookers	Consumer Unit
BR1	1 Pendant	2 Single	–	–	–
BR2	1"	2"	–	–	–
LR1	1"	3"	1 Spur (Switched for Boiler, Pump and Controls)	–	–
Kitchen	1 Batten 1200mm	3 Single 1 Double		30Amp Unit	–
Bathroom	1 Pendant (Pull switch)	–	–	–	–
Lobby	1 Pendant	–	–	–	–
Linen Cupboard	–	–	3kW Immersion with thermostat		
Staircase:					
Landing	1 Pendant (2 way switch)	1 Single	–	–	–
Hall	1 Pendant	–	–	–	–
Distribution	–	–	–	–	1 6 Way Unit

B Wiring to be (PVC) covered 1.5mm^2 for lighting; 2.5mm^2 for socket outlets and immersion heater and 6mm^2 for cooker. Where fixed to walls it shall be chased-in and covered with (PVC) conduit.

C Fittings to be Messrs:

D Provide attendance on installer as described for Heating Engineer, etc.

		Items	Section Totals
		£ p	£ p
Decorations – Throughout			
A	Strip remaining paper from walls and ceilings throughout, clean down, stop cracks and leave ready to receive new decorations.		
B	Line all walls and ceilings and apply (two) coats emulsion in colours to be selected.		
C	Thoroughly clean down existing paintwork, prepare and paint all woodwork (one) undercoat and (one) finishing coat		
D	Prepare and paint (one) undercoat and (one) finishing coat to all exposed pipework.		

	Items		Section Totals	
	£	p	£	p

DRAINAGE, OUTSIDE WORKS, ETC.

Drainage

A Install PVC vent stack, fit balloon guard and joint to drain.

B Install new trapped back/side inlet gulleys, branch drains and manhole in positions, as below to receive rainwater, soil and waste water from fittings including cutting into existing drain. Seal off redundant lengths of old drain in weak concrete:
1 No. Branch drain for vent pipe.
1 No. Branch drain to new WC.
1 No. Branch drain and gulley for bath and basin.
1 No. Branch drain and gulley for sink with side branch and gulley from back addition.
1 No. Manhole on existing drain.

Outside Works, etc

Repairs:

C Take up paved area at rear of house, excavate for and lay 75mm concrete on 100mm hardcore trowelled smooth to falls.

		Items		Section Totals	
		£	p	£	p

GENERALLY

A	Clear rubbish, clean floors and windows and leave ready for occupation.				

Summary

		Section Totals £ p
EXTERNALLY:		
UNDERPINNING	.	
DEMOLITION AND PULLING DOWN	.	
EXTERNAL WALLS	.	
ROOFS	.	
EXTENSION	.	
SUNDRY	.	
PAINTWORK	.	
INTERNALLY:		
STRIPPING OUT	.	
FLOOR	.	
DPC IN EXISTING WALLS	.	
PARTITIONS	.	
FIRST FLOOR:	.	
BR1	.	
2	.	
GROUND FLOOR:	.	
LR1	.	
Kitchen	.	
Bathroom and Lobby	.	
STAIRCASE, HALL AND LANDING	.	
SERVICES:		
COLD WATER SUPPLY	.	
HOT AND COLD WATER	.	
INSTALLATION	.	
HEATING	.	
GAS	.	
ELECTRICS	.	
DECORATIONS:		
DRAINAGE ETC:		
DRAINS	.	
OUTSIDE WORKS, ETC	.	
GENERALLY:		
CLEAR RUBBISH, CLEAN UP, ETC	.	
	SUB TOTAL	
CONTINGENCIES: 5% on Sub Total		
	TOTAL	

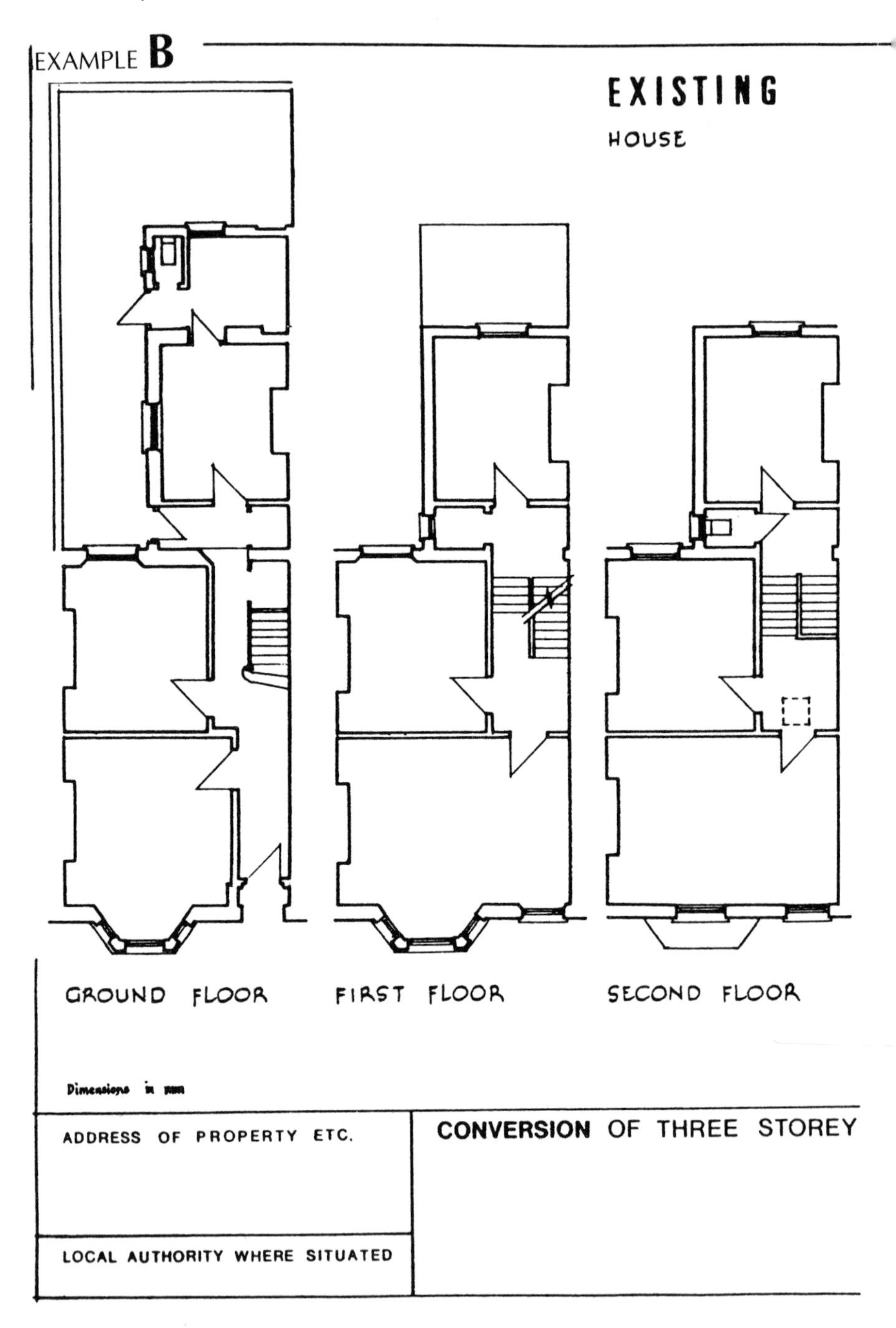
EXAMPLE B
EXISTING
HOUSE
GROUND FLOOR
FIRST FLOOR
SECOND FLOOR
Dimensions in mm
ADDRESS OF PROPERTY ETC.
CONVERSION OF THREE STOREY
LOCAL AUTHORITY WHERE SITUATED

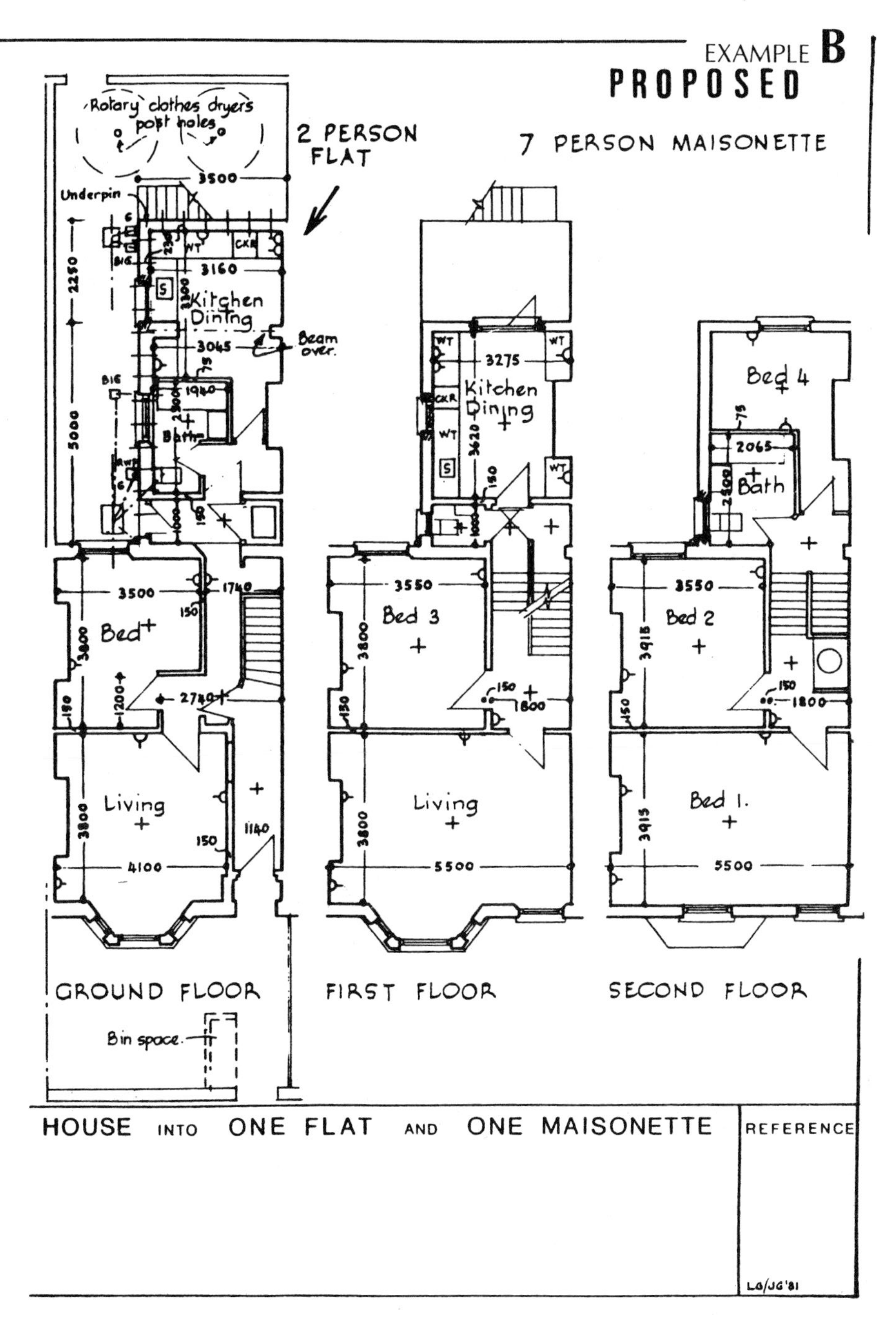
EXAMPLE B
PROPOSED
2 PERSON FLAT
7 PERSON MAISONETTE
Rotary clothes dryers post holes
3500
Underpin
2250
5000
WT
CKR
3160
Kitchen Dining
Beam over.
3045
1940
Bath
Bed
3500
1740
3800
2740
Living
4100
1140
GROUND FLOOR
Bin space
Kitchen Dining
3275
3620
Bed 3
3550
3800
1800
Living
5500
FIRST FLOOR
Bed 4
2065
Bath
Bed 2
3550
3918
1800
Bed 1.
3915
5500
SECOND FLOOR
HOUSE INTO ONE FLAT AND ONE MAISONETTE
REFERENCE
LG/JG'81

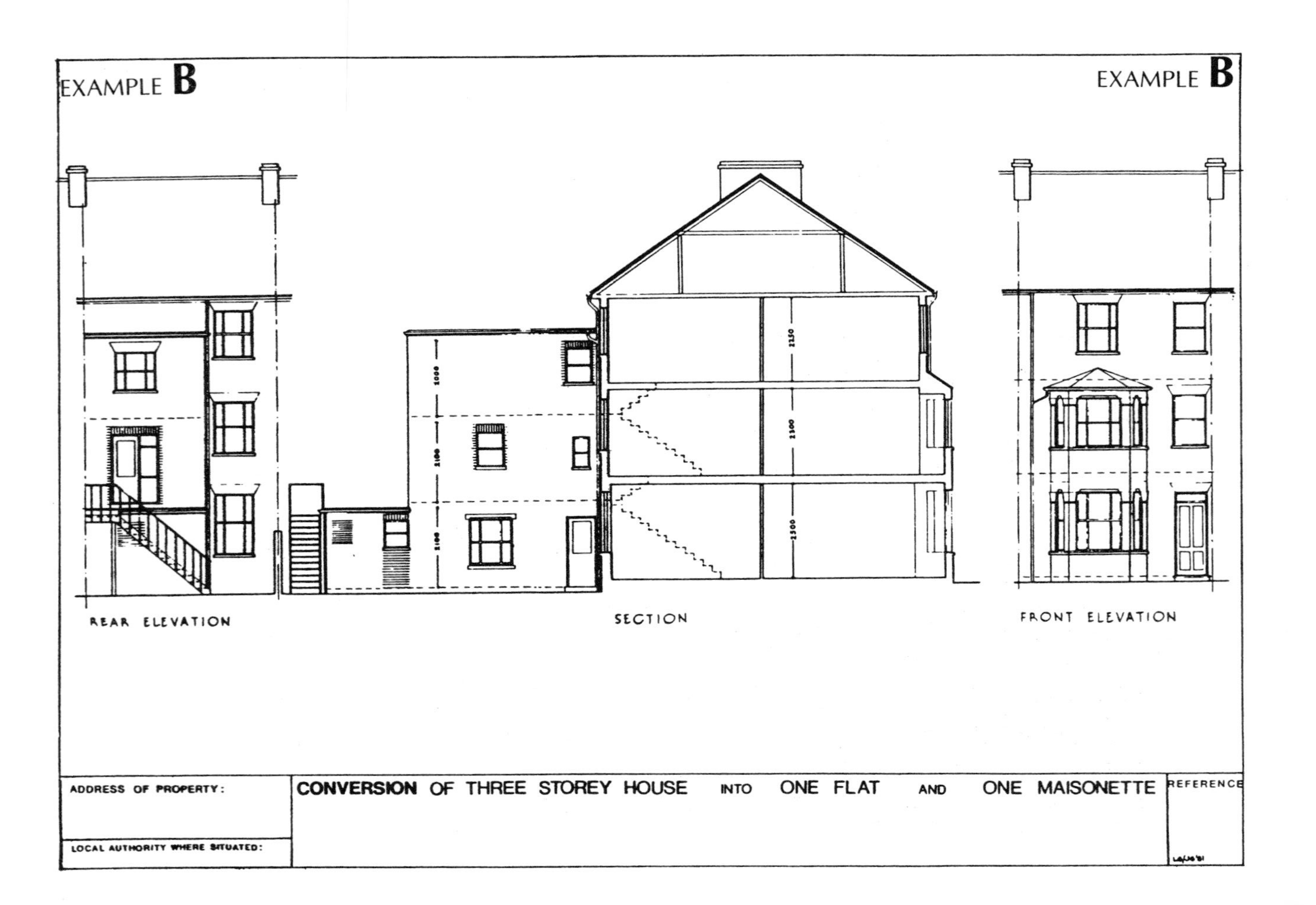
EXAMPLE B
EXAMPLE B
1000
2100
2100
2150
2100
2300
REAR ELEVATION
SECTION
FRONT ELEVATION
ADDRESS OF PROPERTY:
LOCAL AUTHORITY WHERE SITUATED:
CONVERSION OF THREE STOREY HOUSE INTO ONE FLAT AND ONE MAISONETTE
REFERENCE

Example B – Schedule of Works

		Items £ p	Section Totals £ p

EXTERNALLY

Underpinning

A Underpin the external flank wall 343mm thick 5.00m long and 228mm thick 2.25m long to a firm foundation – estimated at not exceeding 1.25m deep below existing paving level.

B Ditto the 228mm thick rear wall to the back addition 3.50m long to a firm foundation – estimated at not exceeding 1.75mm deep do.

External Walls

Isolated Openings

C Take out 2nd floor landing WC window and enlarge opening for proposed bathroom to receive window size 641mm × 920mm.

D Form opening for 1st floor back room window or receive window size 641mm × 920mm.

E Take out 1st floor back room window and enlarge opening to receive composite door and window size 1200mm × 2100mm.

F Take out ground floor back addition WC door and adapt opening to receive window size 641mm × 768mm. Brick-up apron.

For windows and doors – see Internally.

	Items £ p	Section Totals £ p

Isolated Openings continued

A Take out window from ground floor back addition WC and brick-up opening.

B Take out window from ground floor back addition rear wall and brick-up opening.

C Brick-up former door opening in ground floor back addition rear wall presently closed up for about half thickness.

Repairs:

D Repoint both external brick walls including reveals of back addition. Include for replacing approximately 25 No. damaged facing bricks.

E Ditto rear elevation of main building. Include for replacing approximately 10 No. do.

F Ditto front elevation of main building. Include do.

Roofs

Roof Coverings:

G Renew slated covering to slopes of main roof on new battens and felt.

H Renew flat roofs to rear extension with 19mm chipboard and built-up felt roofing.

	Items	Section Totals
	£ p	£ p
Roof Construction:		
A Renew 6 No. roof joists to back addition flat roof.		
Eaves, Gutters and Downpipes		
B Renew gutters to both eaves of main roof with 100mm PVC. Renew all downpipes with 75mm PVC.		
Repairs:		
C Repoint and reflaunch all chimney stacks.		
D Overhaul gutters and downpipes to back addition and renew defective lengths and brackets.		
Insulation:		
E Lay 75mm fibreglass quilt over ceiling joists of main roof and between joists of back addition roof.		
Sundry		
F Fit terra cotta ventilated weatherguards to all discontinued flues.		
Paintwork		
G Prepare and paint (one) undercoat and (one) gloss finishing coat to all existing wood and ironwork.		
H Prepare and paint (two) undercoats and gloss finishing coat to new woodwork.		
I Prepare and apply (two) coats of approved cement paint to all stone cills.		

	Items		Section Totals	
	£	p	£	p

INTERNALLY

A Note: For location of service outlets (eg heaters, electric points, etc) see Gas and Electrical Installations at the end of this Schedule of Works.

Stripping out

B Take out fitments, shelving, battens, curtain tracks, pelmets and other redundant fittings throughout the premises.

C Take down stud partition across the width of the second floor back addition between room, WC and landing.

D Take down the half brick wall enclosing the ground floor back addition WC.

E Take off door to ground floor back room to leave blank opening to new Lobby.

Internal Walls

Isolated openings

F Take out door and frame to ground floor front room and block opening.

G Form opening size 840mm × 2400mm to receive door and frame with fanlight in partition wall between ground floor front and back rooms.

H Take out door and frame from one and half thick brick wall across ground floor back addition back room and enlarge opening to 2000mm × 2400mm.

		Items		Section Totals
		£ p		£ p
Damp Proof Course in Existing Walls				
A	Install damp proof course to all external walls and brick partitions walls by *method to be stated* *(eg silicone injection; insertion of slates; etc).*			
Partitions				
B	Build stud partition to full storey height to form second floor bathroom to overall length of 4.5m with plasterboard covering both sides including forming opening for door with fanlight.			
C	Ditto to divide bedroom 4 from landing to overall length of 1.25m ditto.			
D	Ditto to form second floor airing cupboard full storey height with plasterboard covering both sides including forming opening for door.			
E	Ditto at first floor level to fill spandril between stairs, landing and ceiling × full storey height to form self-contained dwelling with plasterboard covering both sides including forming opening for door with fanlight.			
F	Ditto to form ground floor bathroom to overall length and full storey height with plasterboard covering both sides including forming opening for door with fanlight.			
G	Ditto to form ground floor lobby to living room and bedroom to overall length and full storey height including forming opening for door with fanlight.			

	Items		Section Totals	
	£	p	£	p

Partitions continued

A Ditto to ground floor staircase area to fill spandril between stairs, hall and ceiling × full storey height to form self-contained dwelling and cupboard under stairs to overall length covering one side complete and the other side above the staircase with plasterboard including forming openings for cupboard door and entrance door with fanlight.

Maisonette

Second Floor

BR1:

B Block-in fireplace opening.

Repairs:

C Renew plaster to front external wall.

D Renew plaster to ceiling.

E Piece-in damaged lining and renew door to match existing with mortice lock latch and SAA lever furniture. Reglaze fanlight with 4mm sheet glass with putty.

F Renew skirting to front external wall.

BR2:

G Block-in fireplace opening.

Repairs:

H Renew plaster to rear external wall.

		Items		Section Totals	
		£	p	£	p
Repairs continued					
A	Renew ceiling plaster.				
B	Overhaul double hung sash window.				
C	Ditto door				
D	Renew skirting to rear external wall.				
BR4:					
E	Block-in fireplace opening.				
F	Scrim joints and skim coat plaster to stud partition.				
G	Fit lining for and hang hardboard faced both sides internal door size 686mm × 1981mm on steel butts and fit with mortice lock latch and SAA lever furniture. Glaze fanlight with 4mm sheet glass with putty.				
H	Run wood skirtings to stud partition to match existing.				
Repairs:					
I	Renew ceiling plaster.				
Bathroom					
J	Scrim joints and skim coat plaster to stud partition.				
K	Fit softwood double hung sash window size 641mm × 920mm with hardwood cill, window board, spring balanced, sash lifts and sash fastener and glaze with Group 2 obscured tempered glass with putty.				

	Items		Section Totals	
	£	p	£	p

Bathroom continued

A Fit lining for and hang door. etc as described for BR4.

B Take off remaining skirting and run 12mm × 75mm skirting to all walls and make good plaster where skirting removed.

Install the following sanitary fittings from Messrs range in (white):

C Bath
D Basin
E WC suite

} *see Example "A"*

Sundry:

F Run (two) course (white) tile splashback to one end and one side of bath.

G Ditto to back to Basin.

H Cut opening through external wall at high level and build in air brick externally and vent internally.

Repairs:

I Renew ceiling plaster.

First Floor

LR:

J Block-in fireplace opening.

	Items		Section Totals	
	£	p	£	p
Repairs:				
A Overhaul double hung sash window.				
B Ditto door.				
BR3:				
C Block-in fireplace opening.				
Repairs:				
D Overhaul double hung sash window.				
E Ditto door.				
Existing WC:				
Install basin as described for 2nd floor bathroom.				
Sundry:				
F Run splashback to basin as for 2nd floor bathroom.				
G Cut opening and build-in air brick and vent as for ditto.				
Repairs:				
H Overhaul double hung sash window.				
I Ditto door.				
J Renew WC suite with type as described for 2nd floor bathroom.				

		Items £ p	Section Totals £ p
Kitchen/Dining			
A	Block-in fireplace opening.		
B	Fit softwood double hung sash window size 641mm × 920mm as described for 2nd floor bathroom.		
C	Fit composite door and sidelight to fill opening 1200mm × 2100mm complete with frame and hardwood threshold. Hang door on steel butts and fit with mortice lock latch and SAA lever furniture. Glaze lower panes of door and sidelight with Georgian wired rough cast glass and upper panes with 4mm clear sheet glass. Glaze in putty and with beads to door.		
	Install the following kitchen fittings from Messrs range:		
D	Sink Unit		
E	Floor Units		
F	Worktops — *see Example "A"* (D–H)		
G	Broom Cupboard		
H	Wall Cupboards		
	Install from Messrs range:		
I	Combined stainless steel sink and drainer 1400mm long with plug and chain, Hot and Cold pillar taps connected with copper tails to services, plastic overflow, deep seal trap and waste pipe jointed to stack branch externally.		

	Items		Section Totals	
	£	p	£	p

Kitchen/Dining continued

A Tee off from incoming cold water supply and run extension to point under new sink.

Sundries:

B Cut opening through external wall 300mm × 300mm at high level and build in cast iron air grid externally and aluminium louvred air vent internally.

C Cut out upper panel of existing door, fit beads and glaze with Group 2 obscured tempered glass in putty, rehang door on steel rising butts and renew lock latch and furniture.

Repairs:

D Renew length of skirting to chimney breast.

Flat

Ground Floor

LR:

E Block-in fireplace opening.

F Fit lining for and hang 35mm hardboard faced internal door size 762mm × 1981mm on steel butts and fit with mortice latch and SAA lever furniture. Glaze fanlight with 4mm sheet glass with putty.

		Items £ p	Section Totals £ p
Sundry:			
A	Make out plaster to blocked-up door opening.		
B	Ditto skirting.		
BR:			
C	Block-in fireplace opening.		
D	Scrim joints and skim coat plaster to stud partition.		
E	Fit lining and hang door, etc as described for LR.		
F	Run wood skirting to stud partition to match existing.		
Sundry:			
G	Make out plaster to blocked-up opening.		
H	Ditto skirting.		
Cylinder Cupboard			
Repairs:			
I	Overhaul door.		
Bathroom			
J	Scrim joints and skim coat plaster to stud partition.		
K	Fit lining for and hang door, etc as described for LR.		

		Items £ p	Section Totals £ p
Bathroom continued			
A	Take off remaining skirting and run 12mm × 75mm skirting to all walls and make good plaster where skirting removed.		
	Install the following sanitary fittings from Messrs range in (white)		
B	Bath — As for Maisonette but with waste pipes jointed to gulley inlets externally and WC trap jointed to drain.		
C	Basin		
D	WC suite		
Sundry:			
E	Run splashback to Bath as for Maisonette.		
F	Ditto Basin do.		
G	Cut opening through external wall and build in air-brick and air vent as described for Bathroom in Maisonette.		
H	Reglaze window in Group 2 obscured tempered glass and overhaul.		
Kitchen/Dining			
I	Scrim joints and skim coat plaster to stud partition.		

		Items		Section Totals	
		£	p	£	p

Kitchen/Dining continued

A Fit softwood double hung sash window size 641mm × 768mm with hardwood cill, window board, spring balances, sash lifts and sash fastener and glaze with 4mm glass with putty.

B Fit door for and hang door, etc as described for LR.

C Run wood skirtings to stud partition to match existing.

Install the following kitchen fittings from Messrs range:

D Sink Unit 1400mm long

E Floor Units

F Worktops 1800mm long

G Broom Cupboard

H Wall Cupboards

(D–H: *see Example "A"*)

and from Messrs range:

I Combined stainless steel sink and drainer 1400mm long as for Maisonette but with waste pipe jointed to inlet of gulley externally.

J Tee off from incoming cold water supply and run extension to point under new sink.

		Items £ p	Section Totals £ p
Sundry:			
A	Cut opening through external wall at high level and build in air brick externally and air vent internally.		
B	Make out plaster to blocked up window openings and apron under new window.		
C	Make out skirting to apron under new window.		
Staircase, Landings, Entrance Halls and Lobbies			
D	Scrim joints and skim coat all exposed sides of stud partitions including the inside of cylinder cupboard.		
E	Fit lining for and hang entrance doors to both dwellings half hour fire check internal doors size 762mm × 1981mm on steel butts and fit with mortice lock and 2 No. 150mm SAA bolts.		
F	Fit lining for and hang hardboard faced both sides internal door size 686mm × 1981mm to 2nd floor linen cupboard on steel butts and fit with bales catch and SAA knob furniture.		
G	Fit two slatted wood shelves full depth to linen cupboard.		
H	Run wood skirtings to stud partitions to match existing.		
Sundries:			
I	Make out plaster to blocked up openings.		

		Items £ p	Section Totals £ p
A	Make out skirtings.		
B	Cut out upper panel of existing door to ground floor back addition lobby, fit beads and glaze with Group 2 obscured tempered glass in putty, rehang door on steel rising butts and renew lock latch and furniture.		
C	Piece-in frame to blank opening to ground floor LR/BR lobby.		
Repairs:			
D	Renew plaster to all walls and partition walls affected by rising damp.		
E	Renew ceiling plaster to 2nd floor top landing.		
F	Repair floor boarding to 2nd floor top landing and renew unsound supporting joists and treat with wood preservative.		
G	Renew frame with hardwood cill and overhaul and rehang street entrance door on new steel butts and fit new mortice lock and letter plate and reglaze with Group 2 obscured tempered glass with beads in putty.		
H	Ditto rear entrance door on new steel butts and fit new mortice lock and 2 No. 150mm SAA bolts and reglaze ditto.		

		Items		Section Totals	
		£	p	£	p
Repairs continued					
A	Renew ceiling trap door at 2nd floor with 3mm hardboard faced panel size 500mm × 500mm.				
B	Overhaul staircase and renew treads, handrail and balusters to lowest flight.				

SERVICES	Items		Section Totals	
	£	p	£	p

Hot and Cold Water Installation

Maisonette – First and Second Floors

A Divert existing incoming (lead) cold water supply to the kitchen on the first floor in copper, tee off for sink, continue main in copper to roof space and connect with CWS tank.

B Install on bearers in roof space 50 gallon CWS tank with ball valve and plastic overflow run to point externally at eaves.

C Run Cold down supplies in copper from CWS tank to bath, basins and WCs on first and second floors and to cylinder in airing cupboard on top landing.

D Install in airing cupboard 30 gallon indirect HW cylinder with boss for immersion heater and run copper expansion pipe to discharge over CWS tank.

E Run Hot down supplies in copper to bath, basins and sink on first and second floors.

F Run flow and return in copper from circulator (see Gas Installation) in kitchen to HW cylinder on top landing.

G Install insulation jackets to CWS tank and HW cylinder and insulate all pipes in roof space and under floors.

H Install isolating stop cock on Cold down service.

	Items		Section Totals	
	£	p	£	p

Maisonette – First and Second Floors continued

A Immersion heater – see Electrical Schedule.

Flat – Ground Floor

B Tee off existing incoming (lead) main and install stop cock, tee off for kitchen sink and install stop cock under sink, extend main in copper and connect to Harcopak in "Cylinder Cupboard".

C Install Harcopak Model 50/25/3 direct instant plumbing unit and run overflow in (PVC) to discharge externally.

D Run Cold supplies in copper from Harcopak to bath, basin and WC.

E Run Hot supplies in copper from Harcopak to bath, basin and sink.

F Insulate all pipes under floors.

G Install isolating stopcock on Cold down service.

H Immersion heater – see Electrical Schedule.

Heating

I Heating is by independent room heaters (assumed for this example to be tenants fittings) for which gas or electric points are provided (see Gas and Electrical Installations).

	Items £ p	Section Totals £ p

Gas Installation

A Arrange with Gas Board or their approved installer to provide points for heating and hot water including provision of circulator as follows:

Maisonette – First Floor

B Living Room (near fireplace) : Gas fire point

C Kitchen/Diner (wall mounting) : Circulator point including provision and fixing of circulator with balanced flue

Flat – Ground Floor

D Living Room (near fireplace) : Gas fire point.

E Provide attendance on installer, arrange for connection to incoming main and for metering. Ensure satisfactory completion in sound and working order and pay charges.

	Items		Section Totals	
	£	p	£	p

Electrical Installation

A Renew the electrical installation to the following requirements:

	Lighting	Switch Socket Outlets	Heaters	Cookers	Consumer Unit
Maisonette –					
Second Floor:					
BR1	1 Pendant	3 Single			
"	1 "	2 "			
"	1 "	2 "			
Bathroom	1 " (pull switch)				
Landings	2 Pendants (2 way switch)	1 "			
Airing Cupboard			3kW Immersion with thermostat		
First Floor:					
LR	1 Pendant	3 Single			
BR3	1 "	2 "			
WC	1 "				
Kitchen/ Dining	1 Batten (1200mm)	3 " 1 Double		30Amp Unit	
Landing	3 Pendants (2 way switch)	1 Single			
Flat –					
Ground Floor:					
LR	1 Pendant	3 Single			
BR	1 "	2 "			
Bathroom	1 " (pull switch)				
Kitchen/ Dining	1 Batten (1200mm)	3 Single 1 Double		30Amp Unit	
Cylinder Cupboard			3kW Immersion with thermostat		
Hall (Flat)	2 Pendant (2 way switch)	1 Single			
Hall (Common)	1 Pendant (2 way switch)	1 Single			

			Items £ p	Section Totals £ p

	Lighting	Switch Socket Outlets	Heaters	Cookers	Consumer Unit
Distribution:					
Maisonette					1 unit
Flat					1 unit
Landlord					1 unit

B Wiring:
C Fittings:
D Attendance:
} *see Example "A"*

		Items	Section Totals
		£ p	£ p
Decorations – throughout			
A	Strip remaining paper from walls and ceilings throughout, clean down, stop cracks and leave ready to receive new decorations.		
B	Line all walls and ceilings and apply (two) coats emulsion in colours to be selected.		
C	Thoroughly clean down existing paintwork, prepare and paint all woodwork (one) undercoat and (one) finishing coat.		
D	Prepare and paint (one) undercoat and (one) finishing coat to all exposed pipework.		

		Items	Section Totals
		£ p	£ p

DRAINAGE, OUTSIDE WORKS, ETC

Drainage

A Install PVC soil and vent stack with branches for fittings, fit balloon guard make joints to fittings and drain.

B Install new trapped back/side inlet gulleys in positions to receive rainwater soil and waste water from fittings, as below including cutting into existing. Seal off redundant drains with concrete:

- 1 No. Branch drain for soil and vent pipe.
- 2 No. Branch drains and yard gulley.
- 1 No. Branch drain for GF WC.
- 1 No. Branch drain to GF bath and basin.
- 1 No. Branch drain and gulley with side branch for GF sink.
- 2 No. Manholes on existing drain.

Outside Works

External Staircase – Maisonette to Garden

C Install straight flight bitumen coated steel staircase from flat roof at first floor level at rear 900mm wide rising 2.80m having top landing, handrail to open side of flight and to two sides of landing including bolting to end wall of building and to concrete anchorages formed integrally with concrete paving (see Paving).

		Items	Section Totals
		£ p	£ p
Paving			
A	Take up paved area to side and rear of back addition, excavate for and lay 75mm concrete on 100mm hard-core trowelled smooth to falls, leave post holes for clothes dryers and thicken out to firm ground for bolts for staircase anchorages and cast-in.		
	GENERALLY		
B	Clear rubbish, clean floors and windows and leave ready for occupation.		

	Section Totals £ p
Summary	
EXTERNALLY:	
UNDERPINNING .	
EXTERNAL WALLS .	
ROOFS .	
PAINTWORK .	
INTERNALLY:	
STRIPPING OUT .	
INTERNAL WALLS .	
DPC IN EXISTING WALLS .	
PARTITIONS .	
MAISONETTE – SECOND FLOOR:	
BR1 .	
2 .	
4 .	
Bathroom .	
– FIRST FLOOR:	
BR3 .	
Existing WC .	
Kitchen/Dining .	
FLAT – GROUND FLOOR:	
LR .	
Bedroom .	
Cylinder Cupboard .	
Bathroom .	
Kitchen/Dining .	
STAIRCASE, LANDINGS, ENTRANCE HALLS AND LOBBIES .	
SERVICES:	
HOT AND COLD WATER INSTALLATION .	
HEATING .	
GAS .	
ELECTRICS .	
SUB TOTAL carried forward £	

Section Totals
£ p

SUB TOTAL brought forward

DECORATIONS .

DRAINAGE, OUTSIDE WORKS, ETC
DRAINAGE .
OUTSIDE WORKS, ETC .

GENERALLY:
CLEAR RUBBISH, CLEAN UP, ETC

CONTINGENCIES: The following provision should be inserted for Dayworks:

Craftsmen Labourers Plant and Materials O/heads, Profit, etc %	*Specifier to insert estimated number of hours and amounts for plant and material etc to approximate to a total when priced by the builder as a contingency provision.*

Total

EXAMPLE OF FEINT RULED PAPER FOR SCHEDULE OF WORKS **Fig 1**

Address: (Road and District)							MASTER SCHEDULE		
House Nos: ▷	1	3	5	2	4		Totals	Rate	£
ROOFS:									
Eaves, gutters etc:									
Renew eaves, gutters in 100mm PVC and down pipes in 75mm PVC to front of main building.									
	✓	–	✓	–	–		2		
All as last to both front and rear of main building.									
	–	–	–	✓	✓		2		
Repairs:									
Renew length of eaves gutter where broken matching cast iron to front of main building.									
	–	✓	–	–	–		1		

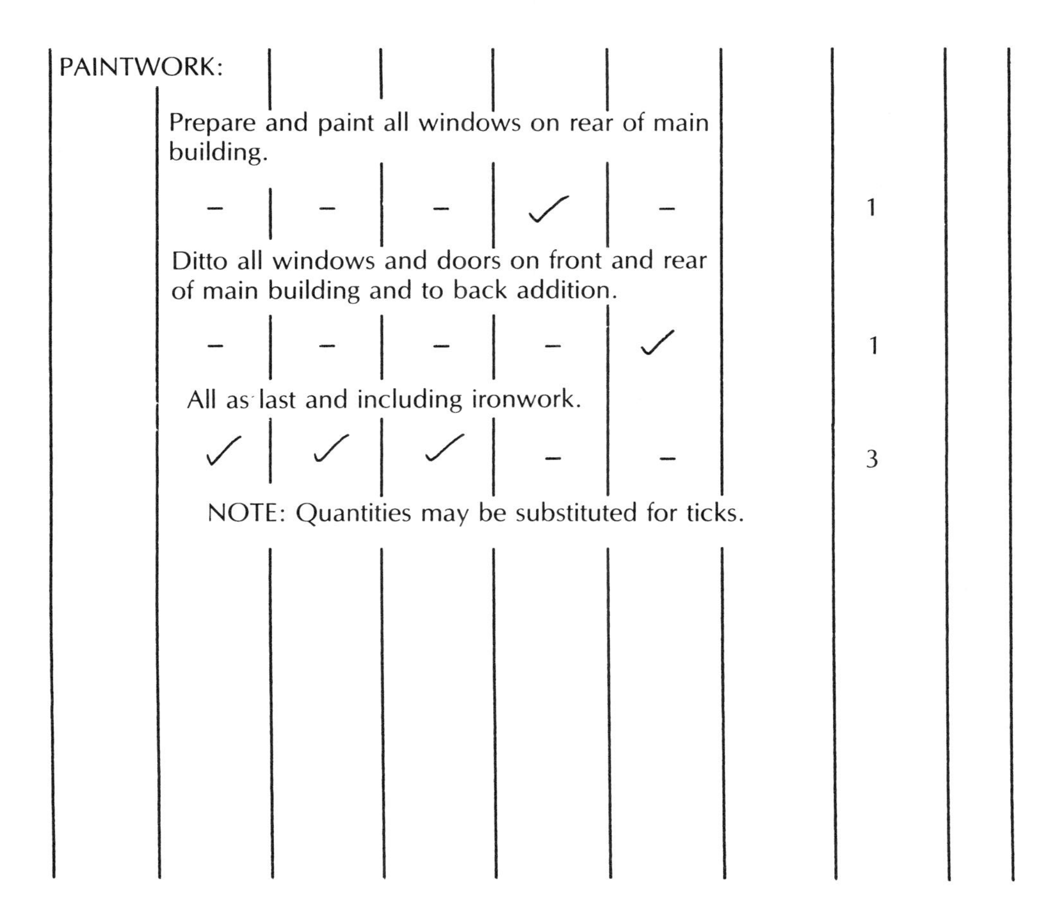

PAINTWORK:								
	Prepare and paint all windows on rear of main building.							
	–	–	–	✓	–	1		
	Ditto all windows and doors on front and rear of main building and to back addition.							
	–	–	–	–	✓	1		
	All as last and including ironwork.							
	✓	✓	✓	–	–	3		
	NOTE: Quantities may be substituted for ticks.							

Section Two

Preliminaries
and
General Matters,
Materials, Workmanship
and
Usage of Sundry Building Terms

Preliminaries and general matters

NOTE: MATERIALS AND PRACTICES IN THIS SECTION ARE INCLUDED AS CURRENTLY ACCEPTABLE STANDARDS FOR THE GENERALITY OF "MINOR WORKS". EQUAL OR BETTER ALTERNATIVES SHOULD BE DEFINED IF REQUIRED FOR ANY PROJECT

2.01 Preliminary Particulars, etc:

.01 Employer

Specifier to state name and address

.02 Architect/Supervising Officer

Specifier to state name and address

.03 General description of the work

Specifier to describe the work and if to be executed in occupied premises, or not

.04 Location and access to the site

Specifier to describe the location of the site and access arrangements

The builder will be responsible for making all necessary access arrangements, negotiations etc with the Local Authority and/or adjoining owners and ensuring due notification to consultants, etc to maintain continuous progress of the works.

2.02 General Matters

.01 Limits of site operations

The builder shall confine his operations to the premises and grounds of the property. The use of a front garden area shall be conditional upon the builder making the necessary arrangements, negotiations etc with the Local Authority and/or adjoining occupiers.

The builder shall prevent workmen, including those employed by sub-contractors, from trespassing on adjoining property.

.02 Preliminary investigations

The builder shall be deemed to have visited and inspected the site and to have examined the drawings and contract documents before pricing and to have adequately acquainted himself with local conditions, accessibility of the works and site, the nature of the ground and sub-soil, the supply of and conditions affecting labour, the availability and supply of materials, water, electricity and telephones, all in relation to the execution of the works as no claim on the grounds of want to knowledge in such respect will be entertained.

.03 Abbreviations and references

The following abbreviations, references or terms are used in this specification:

CP British Standard Code of Practice

BS British Standard Specification

The term "Builder" used throughout this document means the "Contractor" who has contracted to carry out and complete the works.

The reference to "Archt/SO" means Architect, Supervising Officer or other consultant, when appointed, with authority under the terms of the contract to act in such a capacity.

The reference to "Building Inspector" means Building Control Officer, District Surveyor or any other person with authority under statute to require compliance with any relevant regulation or by-law.

The terms "Approved", "Selected" or "Directed" mean at the approval, direction or selection of the building Client/Archt/SO.

The term "Provide" means that the item(s) is to be supplied, delivered and fixed at the expense of the builder.

The term "Daywork" is applicable to work of an unknown or unforeseen nature which could not be specified and therefore estimated for during the preparation of the schedule of works. Such work is usually paid for on a time and material basis with the builder submitting evidence of hours worked and invoices for materials, plant, etc.

Further detailed definitions of terms are contained in "Usage of Sundry Building Terms" at the end of this section and in the Glossary of Terms at the end of section 3.

.04 Prime Cost (PC) and Provisional Sums

PC Sums for the supply of materials

Where PC sums are included in the specification for the supply of materials they shall be deemed to include for delivery to site and to allow a 5% discount for prompt payment on normal monthly account trading terms. Allow for profit if required on such sums and for fixing in position as required.

The term "Fix only" in relation to materials so supplied shall be deemed to include for taking delivery, unloading, storing, moving to position and fixing as required.

PC Sums for sub-contract works

Where PC Sums are included in the specification for the execution of work by a sub-contractor such sums shall be deemed to allow a 2½% discount for prompt payment as required by formal contract conditions.

Allow for profit if required on such sums.

Allow for all attendance upon such sub-contract works in accordance with the custom of the trade concerned but in any case including without charge

provision of standing scaffolding

provision of water, lighting and power

storage of materials

use of any sanitary accommodation

clearance of rubbish

assistance with unloading plant and materials.

Provisional Sums

Such sums, where included in the specification including Contingency Sums, are deemed to include for all necessary labour, materials, plant, overheads and profit and are to be spent only as directed and deducted if not used.

.05 *Contingency Sums (or Dayworks)*

Allow a sum based on a percentage of the value of the works (usually) 5%; alternatively, specify the sum to be provided for Daywork, based on time, materials, plant, etc to enable a competitive rate for daywork to be obtained.

.06 *Pricing Instructions*

The successful tenderer should be prepared to submit a fully priced and monied out specification with the "Schedule of Work" priced out in detail and the total carried forward to a Summary of Building Costs. The Preliminaries, Work Sections, Plumbing, Heating, Gas, and Electrical Installations etc should be totalled separately and their totals inserted in the Summary to Building Costs and unless otherwise instructed, the tender will be deemed to be firm, ie that all allowances for fluctuations in the prices of labour and materials have been taken into account.

.07 *Contract drawings*

Specifier to list

.08 *Form of Agreement and Conditions of Contract*

Specifier to state the Form of Agreement to be used noting any modifications or alterations to standard forms, and listing any information required to be inserted or given in appendices to such standard forms. It is not considered necessary to repeat all the clause headings of standard forms.

SEE EXAMPLES "A" & "B" (Section 1)

Matters requiring to be covered such as Insurance, Payments, Commencement and Completion Dates, Maintenance Period after Completion, VAT, Statutory Deduction Scheme (Finance No 2 Act 1975 requirements) are included in the various formal contract types appropriate to "minor works". Section 4 "Contract" tabulates the clause headings of contract types currently available.

.09 *Management Costs*

Allow for all on and off site management costs.

.10 Plant, tools and scaffolding

Provide all plant, tools, scaffolding or the like necessary for the purpose of proper execution of the works.

.11 Safety, health and welfare of work people

Comply with all relevant regulations involving provision of adequate messing and sanitary arrangements.

.12 Safeguarding the works

Safeguard the works, materials and plant against damage or theft, including all necessary hoardings, watching and lighting for the security of the works and the protection of the public.

.13 Water for the works

Provide water for use in the works.

.14 Lighting and power for the works

Provide artificial lighting and power for use in the works.

.15 Keeping the works clean

Remove all rubbish arising from the works as it accumulates and leave the premises clean and tidy on completion. Allow for cleaning all windows inside and out before handing over the works.

.16 Setting Out

The builder is responsible for setting out the work and making arrangements for this to be checked by the Client or, if appointed, Archt/SO.

Materials and Workmanship

NOTE 1: ATTENTION IS DRAWN TO BS 8000 "WORKMANSHIP ON BUILDING SITES" INTRODUCED IN 1989. IT COVERS MOST TRADES PARTICULARLY RELEVANT TO MINOR WORKS. THEY ARE IN SEPARATE CODES OF PRACTICE

The different Codes of Practice are as follows:

Part 1 Excavation and filling
2 Concrete work
2.1 Concrete on-site mixing
2.2 2 placing, compaction and curing
3 Masonry (including brick and blockwork)
4 Waterproofing
5 Carpentry, joinery and general fixings
6 Roof slating, tiling and cladding
7 Glazing
8 Plasterboard partitions and dry linings
9 Cement/sand floor screeds and concrete floor toppings
10 Plastering and rendering
11 Wall and floor tiling
11.1 Wall and floor tiling – ceramic tiles, terrazzo tiles and mosaics
11.2 Wall and floor tiling – natural stone
12 Decorative wall coverings and painting
13 Above ground drainage and sanitary appliances
14 Below ground drainage
15 Domestic hot and cold services

BS specifications are available from:

BSI Marketing,
Linford Woods,
Milton Keynes MK14 6LE

NOTE 2: IT WILL BE APPRECIATED THERE ARE CONSTANT DEVELOPMENTS IN SPECIFICATION OF MATERIALS AND WORKMANSHIP SO, WHEREAS EVERY EFFORT HAS BEEN MADE TO BE UP-TO-DATE IN THIS TEXT – AS WELL AS CATER FOR THE MORE COMMON TRADITIONAL USAGES – IT IS INDICATIVE. THERE MUST ALSO BE REGARD OF EEC DIRECTIVES.

Excavation

2.03 Materials

.01 *Hardcore*

Hardcore beds or filling shall be of broken brick, concrete, stone, aggregate or similar material to pass a 100mm mesh and shall be free from mud, clay, chalk, timber and rubbish.

.02 *Blinding*

Blinding shall be clean boiler ashes or approved pit sand.

2.04 Workmanship

.01 *Generally*

The excavations for all foundations and formation levels shall be carried out to the dimensions and levels shown on the drawings, as defined in the schedule of works or as ordered by the Building Inspector/Archt/SO. Should any excavation exceed the said dimensions or levels, then the excess excavation shall be filled in with concrete or filling as required by the Building Inspector/Archt/SO.

.02 *Supporting excavation sides*

The stability of trench sides shall be at the sole risk of the builder and any temporary planking and strutting or other temporary works shall be designed by the builder. Should excavation sides fail the Building Inspector/Archt/SO may require the additional space to be filled with concrete or hardcore.

.03 *Finished formations*

The excavations for foundations shall be left open until inspected by the Building inspector/Archt/SO. Final levelling and trimming shall be carried out by hand just prior to pouring concrete foundations.

.04 Hardcore beds or filling

Hardcore is to be deposited in the required thickness of layers, well rammed, consolidated, watered if necessary and blinded.

.05 Disposal of Water

Dispose of all water occurring in the excavations.

Concrete work

2.05 Materials

.01 Cement

Cement shall be ordinary Portland Cement to BS 12. Cement for use in foundations where sulphates exist in the soil shall be Sulphate Resisting Portland Cement as BS 4027.

.02 Aggregates

Fine aggregate for reinforced concrete is to comply with the requirements of BS 882 but Zone 4 aggregate shall not be used.

Coarse aggregate for concrete shall comply with BS 882.

.03 Reinforcement

Mild steel for reinforcement shall comply with BS 4449 and mild steel mesh fabric to BS 4483.

.04 Water

Water shall be clean and free from all harmful matter.

.05 Proprietary concrete lintels

Precast proprietary concrete lintels shall be structurally adequate for the loads to be taken and shall be selected to meet the requirements of the Building Inspector/Archt/SO.

.06 Damp proof membranes

Damp proof membranes shall be polythene sheeting of a minimum of 0.25mm (1000 gauge) in thickness.

2.06 Workmanship

.01 Storage of materials

Cement shall be delivered in bags and stored clear of the ground in a dry weathertight place and shall be used in order of delivery.

Aggregates shall be stored in bunds or in separate stockpiles from any other materials on hard clean surfaces.

Reinforcement shall be stored neatly on a hard surface or in racks.

.02 Concrete mixes

Concrete for use in filling or blinding shall be a 1:12 nominal mix by dry volume.

Concrete for use in wall foundations, beds, paths, concrete lintels and beams and the like, if not particularly specified in the schedule shall be a 1:2:4 nominal mix by dry volume.

Design mixes shall be defined in the schedule of works.

.03 Mixing

Ready mixed concrete from an approved source may be used providing it complies with this specification and BS 5328. Water is to be added to source of manufacture and not on site.

Concrete mixed on site may be mixed by hand or machine. Hand mixing must be on a clean hard surface. Machines and mixing surfaces must be washed thoroughly between mixes.

Measurement of aggregates for mixing must be carried out with a proper gauge box, cement shall be measured by weight and for volumetric proportioning of aggregate 50 kg of cement shall be taken as .035 cu metres.

.04 Testing

Slump test apparatus shall be provided and tests carried out in accordance with BS 1881 when required.

.05 Placing

Concrete is to be placed immediately after mixing and thoroughly worked into place. Concrete shall be in its final position not more than 30 minutes after mixing or delivery.

.06 Concreting in cold weather

The builder shall adopt such measures as are ncessary to ensure that at no time during mixing, placing or hardening does the temperature of the concrete fall below 4.4°C.

No concrete shall be placed when the air temperature is below 2.2°C unless precautions to be taken have been approved.

.07 Concreting in hot weather

Concrete shall be kept continuously moist by frequent watering and shall be kept covered by wet sacking or sand.

.08 Reinforcement

Fabric reinforcement shall be lapped 150 mm at sides and ends.

All loose scale, rust and grease is to be removed from reinforcement before concreting.

.09 Construction joints

Whenever placing of concrete to beds and the like is interrupted on no account shall concrete be left at its natural slope, but is to be stopped at a properly constructed stop board. Lintels and beams shall be poured in one continuous operation.

.10 Formwork

Formwork shall be fixed line to line and level and shall be strongly constructed and adequately braced to support the wet concrete and all construction loads.

.11 Lintels and beams

Lintels over door and window openings and structural beams inserted where load bearing walls or partitions have been removed are to be designed by the builder unless otherwise defined in the schedule of works. Reinforced concrete lintels are to have a minimum bearing of 150mm at each end. Proprietary concrete lintels may be used if suitable, but the builder will still be responsible for design. Dead and imposed loadings shall be calculated in accordance with BS 6399 Part 1.

.12 Damp–proof membranes

Damp–proof membranes at perimeters where the next walls shall be turned up 100mm above the top of the concrete floor slab. Joins between sheets shall be made by double welted folds.

.13 Underpinning

See "Underpinning" 2.09/2.10.

Brickwork and blockwork

2.07 Materials

.01 *Generally*

Materials generally shall conform to the requirements of BS 5628 Part 3.

.02 Not used.

.03 *Cement*

Cement shall be ordinary Portland Cement to BS 12 or Sulphate Resisting Portland Cement to BS 4027 where the mortar is likely to be attacked by sulphates in the soil.

.04 *Lime*

Lime shall be hydrated lime to BS 890 and 6473.

.05 *Premixed lime/sand*

Premixed lime/sand for use in gauged mortar shall be to BS 4721.

.06 *Sand*

Sand shall be clean, sharp pit sand to BS 1199/1200 free of loam, dust or organic matter.

.07 *Water*

Water shall be clean and free from all harmful matter.

.08 *Bricks*

Calcium silicate bricks are to comply with BS 187, concrete bricks with BS 6073 and clay bricks with BS 3921. Clay bricks are to be well burnt and free from stones and lime.

All bricks are to be clean, true to shape with sharp arrises, free from flaws and equal to samples to be submitted. No soft bricks or grizzles will be allowed to be used, nor broken bricks or bats except where required for bond.

Bricks for foundation work shall be as special purpose bricks defined in the British Standards.

Bricks for general use shall be Fletton bricks as manufactured by The London Brick Co Ltd or other equal and approved local brick.

.09 Facing Bricks

Bricks for use as facings shall match existing facing bricks. Wherever possible making out shall be in second hand matching facings of similar age to the existing.

Bricks for facing new walls shall be as defined in the schedule of works.

.10 Building blocks

Building blocks for general use shall be precast concrete blocks to BS 6073 Part 1 Class A and shall have a minimum crushing strength of 3.50 N/mm^2.

Building blocks for structural use shall be as above but with a minimum crushing strength of 7.00N/mm^2.

Building blocks for use in cavity walls or where a high insulation value is important shall be "Thermalite". These blocks have an average compressive strength of 3.50N/mm^2 and are not therefore to be used where the structural requirements are greater than this.

.11 Wall ties

Wall ties are to be galvanised mild steel, 200mm long, to BS 1243. Non-ferrous ties are to be used where the local Building Regulations stipulate these are obligatory.

.12 Damp-proof courses

Damp-proof courses, cavity gutters and the like in new walls are to be asbestos based bitumen felt to BS 6398.

.13 Airbricks

Airbricks are to be terra cotta square hole pattern, size 225 × 75mm for ventilating under ground floors, size 225 × 225mm for ventilating larders and 225 × 150mm in rooms.

.14 Chimney pots

Chimney pots for use with sealed up flues shall match existing pots but shall be of the type with a ventilated weather cover to prevent ingress of rain and to allow ventilation.

2.08 Workmanship

.01 Generally

Workmanship shall generally be in accordance with BS 5628 Part 3.

.02 Storage of materials

Cement and lime shall be delivered in bags and stored clear of the ground in a dry weathertight place and shall be used in order of delivery

Sand shall be stored in a bund or in a separate stockpile from any other materials and on a hard clean surface.

"Thermalite" blocks shall be stored in a dry weathertight place.

.03 Mortar

Cement mortar shall be composed of cement and sand in the proportions of one to four by volume.

Gauged mortar shall be composed of cement, lime and sand in the proportions of one, to one to six by volume.

Mortar shall be thoroughly mixed on clean hard surface and thoroughly incorporated with a sufficiency of water added through a fine rose. The surface is to be cleaned after each mix. Alternatively the mortar may be mixed by machine provided the drum is washed out between mixes. The mortar is to be mixed in small quantities from time to time as required.

No mortar which has commenced to set shall be knocked up for re–use.

Brickwork and blockwork in foundations below damp proof course level are to be laid in cement mortar, all other work is to be laid in gauged mortar unless specifically noted otherwise.

The mortar for pointing face work shall match existing.

.04 New brickwork and blockwork

All new brickwork and blockwork is to be set out and built to the respective dimensions, thickness and heights shown on the drawings and/or as defined in the Schedule of Works. No broken bricks or blocks are to be used except where required to maintain the bond.

Walls are to be carried up in a uniform manner no one position being raised more than 900mm above another at one time.

All perpends are to be kept true, square and in facework plumb with the perpends below and above, and joints are to be of uniform thickness. The whole is to be properly bonded and levelled around at floor level and allowance must be made for such selection and wastage on facing bricks as is necessary.

All joints are to be thoroughly flushed up as the work proceeds. All vertical joints on the face of the walls are to be completely filled except where otherwise specified in cavity wall work.

The cavities of hollow walls are to be kept clear of rubbish and mortar droppings by moveable boards or other approved method. In courses immediately above the bottoms of cavities at the end of each run of cavity, openings are to be left to permit raking and cleaning of cavities on completion. Such openings are to be bricked up afterwards uniform with the surrounding work.

Every fourth vertical joint in the external skin of cavity walls in the course immediately above the bottom of cavities and immediately above flashings or cavity gutters is to be left dry and clear of obstruction.

Wall ties are to be built into cavity walls, spaced 900mm apart horizontally and staggered vertically every 450mm and spaced at 300mm centres around openings.

All brickwork is to rise four courses to 300mm.

All bricks are to be wetted before use, blocks are to be kept dry before laying.

.05 Bond

All half brick walls and all block walls are to be built in stretcher bond.

One brick walls in facings or finished fair face on one or both sides, are to be built in two half brick skins in stretcher bond and shall be tied together with galvanised steel wall ties as described for cavity walls.

Making out existing facework shall be in matching bond.

Other than described above English bond shall be used.

.06 *Frost*

No brickwork or blockwork is to be executed in frosty weather or when the temperature is 0°C or below except that the use of an approved anti–freezing liquid in the mortar mix will be permitted provided that in faced work only complete elevations or sections are laid with the use of the liquid.

.07 *Bonding new work to old*

Where new brickwork or blockwork is built up to or against existing it shall be properly toothed and bonded.

Brickwork or blockwork in walls or partitions abutting old work whether at right angles or in the same plane shall be bonded in alternate courses to pockets or toothings cut into the old work.

Half brick walls or 100mm or less blockwork in thickness to old work shall be tied back with headers built into pockets cut into the existing work or alternatively with galvanised iron cramps one per square metre.

.08 *Facing up old walls*

Where brickwork or blockwork (walls, breasts, piers etc), has been pulled down the old work remaining shall be faced up. Work which is to be covered shall be roughly faced up, all holes or recesses being cut out square and whole bricks or half bricks being inserted so that the finished wall face is structurally sound and level enough for the covering to be applied.

Where the old work is to be left fair face then the face of the entire area is to be cut back and refaced in matching bricks or blocks, mortar, pointing and bond as existing.

The new work is to be tied back with headers or wall ties as described in clause .07, according to bond.

.09 *Quoining up jambs and reveals*

Where new openings are formed or existing openings adapted jambs, reveals and the like are to be quoined up with new bricks or blocks as appropriate properly bonded to match existing and cut, toothed and bonded at the junction with old work.

.10 *Making out facework*

Where making out facework is required following the filling of old door and window openings or the like this shall also include the removal and making out of associated arches, lintels, cills, jambs and the like.

In making out of facework particular attention shall be given to bond and any closers required for bond shall as far as possible match the pattern of the existing work. Pointing to facework shall as far as possible match existing with regard to style, material and colour. Any band courses, patterns or other ornamentation in the facework shall be matched up so that there is no trace of the old opening.

.11 *Building in lintels*

Precast concrete, steel or other pre–formed lintels for openings shall have a minimum bearing on each side of the opening of 150mm. Any packing up and levelling under the bearing ends of lintels shall be in non–crushable material, having regard to the weight of the member.

Filling in and making good above and around lintels shall be carried out before needling or shoring is removed.

.12 *Building in frames*

Frames are to be built in with 5 × 25 × 225mm steel cramps bent and fanged at one end, screwed to back of frame and fanged end built in to brick or block courses, every 600mm. Frames are to be bedded in gauged mortar and pointed externally.

.13 *Openings for Airbricks*

These are to be formed the same overall dimensions as the airbrick and are to be rendered all round inside. In cavity walls the cavities are first to be closed on the four sides with slates bedded in mortar. Lintels are to be provided for openings over 225mm wide.

.14 *Repointing*

Repointing old facework shall include raking out joints of all loose mortar and repointing in matching mortar with the same style of pointing. Any broken bricks, old pipe fixings, cemented up holes etc shall be cut out and new bricks inserted. Any efflorescence, paint splashes, moss or other blemishes shall be removed by wire brushing.

.15 Key for plastering, etc

Brick, block and concrete surfaces shall be keyed for plastering or rendering as required by raking out joints, hacking or applying a suitable bonding agent.

.16 Damp proof courses

Damp proof courses shall be provided in the base of walls at ground level and in external cavity walls over lintels and at sides of openings and to chimney stacks. The material shall extend the full width of walls and be lapped at least 150mm at joints and intersections.

.17 Damp proof courses to existing walls

Damp proof courses to existing walls shall be installed by an approved method and to a standard that will enable the building client to be given a 20 year guarantee.

.18 Underpinning

See "Underpinning" 2.09/2.10.

Underpinning

2.09 Materials

.01 *Generally*

Materials shall be as specified in Excavation, Concrete Work and Brickwork and Blockwork sections. Rapid hardening cement if used shall be to BS 12.

2.10 Workmanship

.01 *Generally*

Workmanship shall be as specified in Excavation, Concrete Work and Brickwork and Blockwork sections where applicable.

.02 *Shoring up*

Shoring up and supporting existing structures shall be at the risk of the builder. Any damage caused to the existing structure or finishes by reason of shoring up or by failure to shore up, shall be made good at the expense of the builder.

.03 *Method of Execution*

Underpinning shall be carried out in short sections not exceeding about 1m in length and in an appropriate sequence of bays, for example, by first executing each end section of the work in turn, followed by the centre section, then alternating end to end working towards the centre.

The concrete and/or brickwork in a section shall be allowed to harden and the work wedged and pinned up to the existing foundations to ensure the stability of the wall before other sections are commenced.

.04 *Concrete Work*

Concrete shall be a 1:2:4 nominal mix by dry volume.

Concrete foundations or walls in underpinning to each section shall be finished against properly constructed stop boards with joggle joints for key and not against earth surfaces.

To facilitate wedging and pinning up concrete walls or foundations a small gap shall be left under the existing foundations to be underpinned. When the new concrete walls or foundations have hardened and settled the gap shall be filled with a dry 1:2:4 concrete mix well rammed home.

.05 Brickwork

Bricks for use in underpinning shall be of special quality as defined in BS 3921 and shall have adequate compressive strength for the particular use and situation.

All brickwork shall be built in cement mortar 1:4 and in English bond.

Brickwork to each short section shall be properly set out and toothings left for joining up to adjacent sections, so that when the whole length of underpinning brickwork has been completed it is properly bonded together.

Wedging and pinning up brickwork to the underside of old foundations shall be carried out with stout slates in stiff cement mortar.

Asphalt work

2.11 Materials

.01 Mastic Asphalt

Mastic asphalt shall be composed of either natural rock asphalt or limestone aggregate in accordance with the relevant British Standards as listed below:

	Natural	Limestone
Roofing	BS 6577	BS 6925
Tanking Damp proof courses	BS 6577	BS 6925
Flooring	BS 6577	BS 6925 BS 6925
Roadways & Footways	BS 1446	BS 1447

.02 Sundry materials in connection with asphalt roofing

Generally roofing materials shall be in accordance with CP 144 Part 4.

Isolating membranes for wet or dry construction shall be impregnated flax felt to BS 747 Type 4A.

Vapour barriers for protecting insulating materials shall be coated roofing felt to BS 747 Type 1B.

Metal lathing used for keying shall be plain expanded metal lathing complying to BS 1369.

Flashings for roofing works shall be either lead or preformed aluminium as described in ROOFING.

Chippings for roofing work shall be as defined in CP 144 Part 4.

.03 Sundry materials for use with asphalt flooring

Generally these shall be in accordance with CP 204.

2.12 Workmanship

.01 Generally

Workmanship generally shall be in accordance with any relevant code of practice viz:

Tanking and damp proof courses	CP 102
Roofing	CP 144 Part 4
Flooring	CP 204

.02 Preparation of surfaces

Brickwork and blockwork surfaces shall have the joints raked out to a depth of 10mm and be well brushed down.

Concrete surfaces shall be clean and rough. Smooth surfaces produced by formwork or trowelling are unacceptable and must be hacked for key and afterwards brushed down.

Timber surfaces of upstands, kerbs and the like in roofing work shall be covered with expanded metal lathing securely stapled on.

.03 Coats and thicknesses

Tanking and damp proof courses shall be applied in three coats to a total thickness of 30mm generally, and in three coats to a total thickness of 20mm to vertical surfaces and slopes over 30° to the horizontal. Two coat angle fillets should be applied to junctions of surfaces.

Roof coverings shall be applied in two coats to a total thickness of 20mm generally, and in two coats to a total thickness of 13mm to vertical surfaces and slopes over 30° to the horizontal. Two coat angle fillets should be applied at internal angles formed by the junctions of different surfaces.

Floor finishes shall be applied in the number of coats and to the thickness specified in the schedule.

.04 Finishes

Surface finishes of roofing shall be a bitumen dressing compound with mineral aggregate chippings.

Surface finish of flooring shall be as specified in the schedule.

Roofing

2.13 Materials

.01 Tiles and slates

Tiles and slates used in repairs or extensions shall match existing. Roofs which are to be completely retiled or reslated or new separate roofs shall be in tiles or slates specified in the schedule of work and in accordance with one or more of the following British Standards:

Clay Tiles	BS 402
Concrete Tiles	BS 473
Slates	BS 680
Asbestos slates and sheets	BS 690

.02 Battens

Battens shall be sawn softwood pressure impregnated with preservative to BS 4471 as described in Carpentry and Joinery, size 50 × 25mm for plain tiling and 38 × 19mm for single lap tiling and slating.

.03 Roofing felts

Roofing felts shall be in accordance with BS 747. Sarking felts under tiled and slated roofs shall be type 1A. Built up felt roofing to flat roofs shall comprise three layers of felt, each layer selected in accordance with the requirements of CP 144 Part 3 and finished with self finished felt or surface chippings as recommended. Vapour barriers shall be provided as recommended in CP 144 Part 3 when dry insulation is specified under the roofing.

.04 Lead sheet

Lead sheet for roofing, flashings and coverings shall be to BS 1178.

.05 Zinc sheet

Zinc sheet for roofing, flashings and coverings shall be to BS 6561.

.06 Woodwool decking slabs.

Woodwool decking slabs shall be to BS 1105 type B.

2.14 Workmanship

.01 Generally

The workmanship generally shall comply to the requirements of the various codes of practice for roofing:

Slating and tiling .... BS 5534 Part 1

Sheet roof coverings .... CP 143

Roof coverings .... CP 144

Any manufacturer's instructions or recommendations for any particular material shall be followed.

.02 Plain tiling

Battens shall be spaced to the gauge required and nailed to each rafter with composition nails 70mm long.

Each tile in every fourth course shall be nailed with two stout composition nails, also tiles at the end of courses adjacent to abutments, verges, hips and valleys, and all tiles in courses at eaves and top edges.

Sarking felt shall be laid parallel to ridges and lapped 150mm both ways. A separate strip of felt (roll width) should be laid along ridges, hips and valleys.

Verges shall be formed with a tile undercloak and with tile and a half tiles to maintain bond. Cut tiles are not acceptable. Verges should project beyond gable walls 50 to 75mm and shall be bedded and pointed in cement mortar (1:3).

Lead soakers shall be used on all side abutments together with flashings or tile listings so as to match existing roofs. Edges of tile listings must be cut into the abutting wall. Cement fillets are not to be used on abutments and where this was the existing method of finishing abutments listings and soakers should be used instead. Lead apron flashings shall be used on top edge abutments.

Eaves shall be formed with special under tiles each nailed as previously described. Eaves should overhang gutters by 38 to 50mm.

Ridges, hips and valleys shall be formed to match existing roofs, bedding and pointing in cement mortar if the type of tile so requires. Hip irons, 6 × 25mm section, shall be provided.

.03 Single lap tiling

All the clauses relating to plain tiling shall apply where relevant and except where amended by the following.

Manufacturer's instructions and recommendations shall be followed particularly with regard to the provision of special tiles for verges, top edges, eaves and the like.

Each tile in alternative courses shall be nailed with one nail.

Lead cover flashings shall be used at side abutments.

.04 Slating

Battens shall be spaced to the gauge required and nailed to each rafter with composition nails 70mm long.

Each slate shall be twice centre nailed with composition nails of lengths varying from 63 to 32mm according to size and thickness of slates. Nail holes shall be a minimum of 25mm from edges.

Sarking felt shall be provided as for tiled roofs.

Verges shall be formed with a slate undercloak, project from 50 to 75mm, and be bedded and pointed in cement mortar (1:3).

Lead soakers shall be used on all side abutments together with flashings or tile listings so as to match existing roofs all as described for plain tiling. Lead apron flashings shall be used for top edge abutments.

Eaves shall be formed with a double course of slates head nailed and should overhang gutters 38 to 50mm.

Ridges, hips and valleys shall be formed with lead coverings or special tiles to match existing roofs. Hip irons, 6 × 25mm section shall be provided to tile hips.

.05 Built up felt roofing

Built up felt roofing shall only be used on flat roofs and shall comprise three layers of felt with a chippings finish. Flat roofs shall have a fall of 1 in 80. On timber sub–bases the bottom layer of felt is to be nailed down with galvanised clout headed nails at 300mm centres both ways. On screeded sub–bases and between layers felt shall be fully bonded in bitumen.

Upstands against walls shall be covered with lead flashings. Edge trims shall be in dressed lead or pre–formed aluminium.

Chippings shall be omitted from gutters.

.06 *Lead sheet*

The following weights of lead sheet shall be used for the uses listed:

roofs and main gutters	Code 6
ridges, hips and valleys	Code 5
flashings, damp proof courses and chimney gutters	Code 4
soakers and slates	Code 4

Lead flats and gutters shall have a minimum fall of 1:80. Drips and rolls shall be spaced regularly to suit sheet sizes. Upstands against walls shall be covered with flashings.

.07 *Zinc sheet*

Zinc sheet for roofing, flashings and coverings shall be in zinc gauge 15.

Zinc flats and gutters shall have a minimum fall of 1:80. Drips and rolls shall be spaced regularly to suit sheet sizes. Upstands against walls shall be covered with flashings.

.08 *Decking*

Woodwool decking slabs are to be laid with close dry butt joints and fixed to joists with 100mm galvanized clout nails at 300mm centres. End joints between slabs must always be located over a joist and each slab must have five nails across its width.

Woodwork

2.15 Materials

.01 Timber

All timber shall be well seasoned, bright, sound, cut square and straight grained and shall be free from discoloured sapwood, wane, shakes, dry, loose or dead knots, or any other defects which will render it unsuitable for its intended use.

Timber for carpenter's work shall be in accordance with BS 4978, BS 4471 and BS 5268 and shall have a moisture content of not more than 20% and not less than 15% of the dry weight at the time of fixing. The timber for structural use shall be graded in accordance with BS 4978.

Timber for joiner's work shall be in accordance with BS 1186 part 1 Class 1S for hardwood and clear finished softwood, and Class 2 for softwood which is not concealed. Timber shall be used in accordance with the uses permitted in tables A & B. Timber for flooring shall be graded and sized in accordance with BS 1297 and shall be in accordance with CP 201 Part 2.

.02 Plywood

Plywood shall be in accordance with BS 6566 bonded with MR adhesive to BS 1203 for internal use and WBP adhesive to BS 1203 for external use.

.03 Harboard

Hardboard shall be in accordance with BS 1142 type TN and shall have a flame spread classification of Class 1A.

.04 Wood Chipboard

Wood chipboard shall be in accordance with BS 5669.

.05 Blockboard and laminboard

Blockboard and laminboard shall be in accordance with BS 3444.

.06 Treated timber

Treated timber shall be pressure impregnated with an approved preservative complying with the requirements of BS 3452 in areas subject to insect attack and BS 4072 in conditions subject to fungal decay constructional timber shall be treated in accordance with BS 5268 Part 5. All cut surfaces and notches made on site shall be treated with a suitable brush applied preservative.

.07 Asbestos and Non–Asbestos Insulation Board

Removal of asbestos insulation board shall be in accordance with the recommendations of the Health and Safety Executive and replacements shall meet the methods of test as specified in BS 4624.

.08 Woodwool slabs

Woodwool slabs for non–loadbearing partitions and insulation purposes shall be in accordance with BS 1105 type A.

.09 Glues

Glues shall be in accordance with BS 1203/1204 and the appropriate grade shall be selected according to use and location.

.10 Flush doors

Flush doors for internal and external use shall be plywood faced in accordance with BS 459 Parts 2 and 3. (N.B. Part 2 is subject to amendment in accordance with Draft for Development (DD) 171/1987 "Guide to specifying performance requirements for hinged and pivoted doors including test methods". Part 3 is under review).

.11 Windows

Subject to any requirements to match existing designs or sizes, timber casement windows and double-hung sash windows shall be in accordance with BS 644. Timber sub–frames to metal windows shall be in accordance with BS 1285 with type 2 cills.

.12 Staircases

Subject to any requirements to match existing work, internal staircases and balustrades shall comply with BS 585.

.13 Kitchen Fittings

Kitchen fittings shall be timber in accordance with BS 1195 Part 2 and BS 6222 Part 1. Worktops shall be continuous over the base units to avoid unsightly joins. The body of the fittings shall be sprayed with white enamel paint.

.14 Ironmongery

(Specifier to state requirements.)

2.16 Workmanship

.01 Storage

Timber shall be stacked clear of the ground and protected from the weather.

.02 Priming

All softwood door and window frames and linings are to be delivered to site primed. Priming shall be carried out with the correct primer as specified under painting.

.03 Jointing, fixing and assembly

The workmanship generally shall comply to the requirements of BS 1186 Part 2 and CP 5268.

All framed work shall be cut out and put together immediately upon receipt of the details but shall not be glued and wedged up until ready for immediate fixing.

External joinery shall be put together with a WBO grade adhesive to BS 1204.

Where nails are used for fixing softwood the nails are to be punched in.

All screws are to be countersunk. Screws for fixing hardwood are to be either sunk or pelleted or if exposed are to be brass. Screws for fixing ironmongery to be matching.

Plugging for fixing timbers shall be at 400mm centres unless otherwise described.

Skirtings and the like shall be in single lengths wherever possible and jointed with splayed heading joints otherwise.

.04 Not used.

.05 Studding

Studding to partitions, casings etc shall generally be executed in 75 × 50mm members. There shall be a sole and head plate and vertical studs at not more than 600mm centres. Noggins shall be provided at not less than 900mm centres vertically.

.06 Defective work

Any new joinery that splits, shrinks or warps is to be renewed or replaced without charge.

.07 Repairs to old joinery generally

Broken or damaged members which cannot be repaired by filling and decorating shall be cut out and new sections or timber let in, glued and pinned and shaped or moulded to match the existing member.

.08 Renovating timber boarded floors

The renovating of timber boarded floors shall include removing all linoleum, carpet or other overlays and all tacks and fixings. All floor nails shall be punched in and boards made secure, any missing nails to be replaced. Where joists are too worn to provide adequate fixing for floor nails the worn sections shall be cut out and new timber let in. Any missing, split or otherwise defective boards shall be replaced. Any uneveness between adjacent boards shall be checked and planed off.

.09 Making out timber boarded floors

Where openings formed by hearths, staircases, ducts and the like are to be filled in the junction of new and old boardings is to be staggered every board so that continuous straight joints are avoided. Boarding used for making out must match the existing boarding in width and thickness.

.10 Sub–floors for flexible finishings

Sub–floors for flexible finishings on existing boarded floors shall be formed with 4mm plywood sheets screw nailed to the floor boarding at 150mm centres both ways.

Plasterwork

2.17 Materials

.01 Cement

Cement shall be ordinary Portland Cement to BS 12.

.02 Lime

Lime shall be hydrated lime to BS 890 and BS 6463.

.03 Sand

Sand for internal plastering with gypsum plasters shall comply with BS 1199/1200.

Sand for external renderings, internal plastering with lime and Portland Cement and floor screeds shall comply with BS 1199/1200.

.04 Water

Water shall be clean and free from all harmful matter.

.05 Plasters

Gypsum plasters shall comply with BS 1191 Part 1.

Premixed lightweight plasters shall comply with BS 1191 Part 2.

Choice of suitable material from within the above ranges shall be at the builders option.

.06 Plasterboard

Plasterboard shall comply with BS 1230.

Plasterboard which is to be decorated or receive lining paper shall be self finished gypsum wall board.

Plasterboard which is to receive a setting coat only of plaster shall be gypsum lath with rounded edges.

.07 Expanded metal lathing

Metal lathing for plastering shall be plain or ribbed expanded 22 gauge mesh to BS 1369.

.08 Plaster beads

Metal plaster beads and the like shall be "Expamet" type obtained from The Expanded Metal Co Ltd.

.09 Wall tiles

Wall tiles shall be 108 × 108 × 6mm white glazed cushion edged tiles to BS 6431. Joints shall be grouted in white cement. Rounded edge tiles shall be provided on exposed edges.

.10 Clay floor tiles

Clay floor tiles shall be 152 × 152 × 10mm to BS 6431. Joints shall be grouted in neat cement.

.11 Decorative flooring

PVC (Vinyl) asbestos floor tiles shall be in accordance with BS 3260. Flexible PVC flooring, sheet or tiles, shall be in accordance with BS 3261.

.12 Air Vents

Air vents to rooms shall be fibrous plaster pattern of sizes to suit vent apertures.

.13 Mastic for pointing

Mastic for pointing shall be white polysulphide mastic to BS 4254.

2.18 Workmanship

.01 Generally

Standards of workmanship generally shall comply to BS 5492 for internal plastering and BS 5262 for external rendered finishes.

.02 Arrises and edges

Metal angle beads shall be used on all vertical edges and any horizontal edges within normal reach. Stop beads shall be used on all exposed plaster edges and where plaster abuts other surfaces without being masked by a cover fillet.

.03 *Plastering or rendering to walls and ceilings*

Plastering or rendering generally shall be in two coat work nominal thickness 13mm.

Brick, block and concrete surfaces shall be keyed for plaster or rendering as required by raking out joints, hacking or applying a suitable bonding agent.

Plasterboard backings shall be not less than 9mm thick generally and fixed with galvanised clout nails and the joints between boards covered with jute scrim. Where board joints occur additional timber noggins shall be provided for fixing if required, so that boards are fixed on all four edges.

.04 *Plastering to plasterboard on walls and ceilings*

Plastering to plasterboard backings to walls or ceilings shall be a setting coat only except that additional thickness of plaster is required to bring tapered joints out level.

.05 *Making out and making good to old plastered surfaces*

All loose plaster to solid or metal lathing backings shall be cut back and the surfaces to be plastered keyed as required. All surfaces shall be thoroughly brushed down and wetted before plastering.

Allow for dubbing out in cement and sand as found necessary in order to finish new plaster surfaces level with existing.

Where fireplaces, doors, windows etc have been removed and the openings sealed for plastering, the old plaster is to be cut back until a straight edge can be levelled across the opening in all directions.

.06 *Plastering damp walls*

Walls affected by rising dampness are to have skirtings removed and plaster hacked off the walls up to 450mm above the highest point of the damp plaster as detected by a moisture meter. After being allowed to dry out for the maximum possible time they are to have the joints raked out and replastered with cement and sand (1.3) rendering coat gauged with a proprietary waterproofing/salt retardant liquid as recommended by the manufacturer of the liquid. The floating coat is to be cement and sand (1.4) and the setting coat to be a porous plaster mix (eg lime and "Sirapite" plaster (1.1) and not over trowelled.

.07 Making out and making good to old plasterboard surfaces or backing

Plasterboard backings shall be cut and fitted accurately into the openings to be made out and the joints scrimmed and the whole plastered with the same number of coats as existing.

Where existing openings in self finished plasterboards are to be sealed all existing cut boards around the old opening are to be removed and reinstated with whole new boards.

.08 Beds and backings

Beds and backings for floor and wall finishings shall be in cement and sand mixed in the proportions 1:4 by volume and shall be laid to the thickness required by the schedule of works or the drawings. Where no specific instructions are given beds shall be 40mm thick and backings 13mm thick. Beds shall be finished with a wood float or steel trowel according to the requirements of the floor finish to be received.

.09 Wall tiling

Joints of tiling shall be pointed in white cement. Junctions between tiled splash backs and sanitary appliances or worktops shall be pointed in white mastic.

.10 Floor tiling

Any floor tiling shall be laid in accordance with CP 202 and sheet and tile flooring in accordance with BS 8203.

Steelwork and metalwork

2.19 Materials

.01 Steel beams

Steel members for builder designed beams shall be structurally adequate for the loads to be taken and so as to be visually acceptable.

.02 Proprietary steel lintels

Proprietary steel lintels shall be structurally adequate for the loads to be taken and shall be selected to meet the approval of the Building Inspector/Archt/SO.

.03 Metal windows

Metal windows are to be in accordance with BS 990 and aluminium windows with BS 4873. Sub–frames for steel windows are to be as described in Woodwork.

.04 Balustrades and staircases

Balustrades and staircases shall be constructed in steel and generally in accordance with the requirements of BS 6180 and Section H of the Building Regulations.

2.20 Workmanship

.01 Beams and lintels

Lintels over door and window openings and structural beams inserted where load bearing walls or partitions have been moved are to be designed by the builder unless otherwise defined in the schedule of works. Adequate bearings are to be provided at each end and if required special padstones shall be provided to spread the load in the case of large beams. Proprietary steel lintels may be used if suitable, but the builder will still be responsible for design.

Structural steelwork not encased in concrete is to be primed with two coats of red oxide primer before fixing.

Dead and imposed loadings shall be calculated in accordance with BS 6399 Part 1.

.02 Fixing metal windows

Windows shall be fixed in brick reveals or sub–frames as specified, plumbed in square and levelled.

Metal frames shall be bedded to sub–frames or brickwork in mastic and pointed externally.

Screwed fixings shall be properly plugged and lugs built in securely.

.03 Balustrades and staircases

Balustrades and staircases shall be installed in accordance with the recommendations of BS 6180 as far as this is applicable. When fixed balustrades shall be firm and rigid under hard pressure.

Plumbing and engineering installations

2.21 Materials

.01 Generally

All materials selected must be in accordance with building regulations, by–laws and water authority requirements.

.02 Rainwater goods

Isolated lengths of gutter or pipe to be renewed shall be in materials to match existing. Complete new installations shall be in u PVC cast iron or other – as defined in the schedule of works.

Unplasticised PVC half round rainwater gutters and round pipes and fittings shall comply to BS 4576 Part 1.

Cast iron half round rainwater gutters and round pipes and fittings shall comply to BS 460.

.03 Sanitary plumbing

Main vertical soil, waste and ventilating pipes and fittings shall be in cast iron to BS 416; u PVC to BS 4514 or other – as defined in the schedule of works.

Waste and overflow pipes, fittings and traps shall be in copper; u PVC or other – as defined in schedule of works. Copper tubing shall be to BS 2871 Part 1, and shall be fixed with single copper saddle clips or approved white plastic single screw fixings; fittings for copper tubing may be either capillary or compression type to BS 864 and traps shall be to BS 1184. u PVC pipes and fittings shall be as Key Terrain system 200 or equal, traps shall be in plastic to BS 3943.

Waste fittings for sanitary appliances and overflow fittings for baths shall comply to BS 3380.

.04 Coldwater pipework

Cold water pipework to be laid underground shall be in copper to BS 2871 Part 1 table Y with fittings to BS 864, polythene (type 32) to BS 1972,

6572 and 6730; polythene (type 50) to BS 3284, 6572 and 6730 or other – as defined in the schedule of works.

Copper pipes laid undergounds shall be wrapped in "Denso" tape.

Cold water pipework above ground shall be in copper; u PVC or other – as defined in the schedule of works. Copper tubing shall be to BS 2871 Part 1, and shall be fixed with single copper saddle clips or approved white plastic single screw fixings except where lagging is required when two piece copper spacing clips shall be used; fittings may be either capillary or compression type to BS 864. u PVC tubing shall be to BS 3505 and shall be fixed with approved plastic clips designed to keep pipes clear of walls where lagging is required; fittings shall be to BS 4346.

.05 Hot water pipework

Hot water pipework shall be in copper tubing to BS 2871 Part 1 fixed and with fittings as described for cold water services.

.06 Heating pipework for small bore closed circuit central heating installations

Heating pipework shall be in copper to BS 2871 Part 1 all as described for hot water pipework – as defined in the schedule of works.

.07 Gas pipework

Gas pipework shall be in copper to BS 2871 Part 1 fixed and with fittings as described for waste pipes; in steel tube to BS 1387 where not exposed or other – as defined in the schedule of works.

.08 Water storage

Cold water storage cisterns shall be in polyolefin; olefin copolymer materials to BS 4213 or other – as defined in the schedule of works.

Ball valves shall be to BS 1212 Part 1 and floats in copper to BS 1968; plastic to BS 2456 or other – as defined in the schedule of works.

Hot water storage shall be in copper direct to BS 699; copper indirect to BS 1566; copper combination cylinders to BS 3198 or other – as defined in the schedule of works. Provision for an immersion heater shall be made.

.09 Draw off taps and stopvalves

Draw off taps and stopvalves for hot and cold water services shall be to BS 1010.

.10 Hot water convectors and radiators

Convectors and radiators installed in domestic hot water central heating systems shall comply to BS 3528. Radiator valves and unions shall comply to BS 2767 and to BS 6284, for thermostatic valves. Convectors and/or radiators shall be of modern design, visually acceptable and approved.

.11 Insulation

Pipework to be insulated shall be wrapped with mineral wool or preformed foam pipe lagging adequately secured.

Cold water cisterns and tanks shall be adequately insulated. Hot water cylinders and the like shall be insulated with properly designed plastic covered insulation.

.12 Heat source, circulating pumps and other equipment for hot water and central heating installations

Specifier to state requirements.

.13 Sanitary fittings

Specifier to state requirements.

2.22 Workmanship

.01 Generally

The design and pipe sizing of all domestic heating installations and sanitary, hot and cold water plumbing shall be the responsibility of the builder subject to routing of pipework being approved.

The entire plumbing and engineering installations shall be designed and installed in accordance with good practice and accepted engineering standards, shall accord with any relevant Code of Practice and shall comply with building regulations, water authorities regulations or by–laws. The builder shall carry out such tests as may be required to satisfy these requirements.

.02 Pipework

Wherever possible pipework shall be out of sight under floor boards, in ducts or cupboards and in roof spaces or the like.

Pipework shall be adequately fixed to walls or other surfaces. Care shall be taken in the installation of hot water and heating pipework to avoid vibration being transmitted from circulating pumps.

.03 Stop valves

Stop valves shall be provided on all hot and cold water services in the following positions:

(1) Cold water rising main on entry to the building and before every appliance directly served.

(2) On all cold water down services from a storage cistern at the cistern.

(3) On all hot and cold water service pipes before every appliance, or range of appliances served.

.04 Access to soil and waste pipes

Access doors shall be provided on all main soil and waste stacks with proper builder's work access panels where necessary.

All waste traps shall be accessible.

.05 Insulation

All water storage cisterns, cylinders and the like shall be insulated. All pipework in roof spaces shall be insulated.

.06 Builder's work

All water storage cisterns, tanks, cylinders and the like shall be adequately supported with joists, bearers and boarding etc and pipes shall be supported throughout their length to prevent misalignment.

Holes through existing or new floors, walls and ceilings shall be provided as required and made good whether or not specifically mentioned elsewhere.

All exposed pipework shall be painted as described in Painting and Decorating.

Electrical installations

2.23 Materials

.01 Generally

Materials are to comply with the latest IEE Regulations and any relevant British Standards

.02 Cables, conduits and accessories

The types of cables, conduits and accessories used for wiring must be suitable for the purposes used and must be consistent throughout the installation.

.03 Fittings

Ceiling roses, light switches, socket outlets or spur connectors and the like are to be white plastic of approved pattern or as defined in the schedule of works.

All switches and outlets are to be flush pattern as defined in the schedule of works.

.04 Consumer unit

The consumer unit shall be a neat and compact metal clad unit with mini circuit breakers (not fuses). Sufficient ways should be provided for lighting, ring main power circuits, electric cooker, immersion heater, door bell and spare. (Whether or not all these are installed.)

.05 Immersion heater

Electric immersion heaters shall be 3kW/1kW split element type to BS 3456.

.06 Cooker control units

Cooker control units shall be flush metal cased and clad to BS 4177, finished to match other fittings.

.07 Night storage heaters

Specifier to state requirements.

2.24 Workmanship

.01 Generally

The entire installation is to be designed and installed by the builder to the requirements of this specification and the drawings and/or schedule of works.

The installation is to comply with the latest IEE Regulations and any requirements of the local electricity board. The builder shall carry out such tests as may be required to satisfy these requirements.

.02 Conduit/cable

All conduit/cable shall be hidden, no surface wiring of any sort will be allowed.

Cable buried in plaster if not in a conduited system must be protected by conduit or other suitable covers against accidental penetration from nails and drills.

Conduit if used shall be securely and neatly fixed with proper clips to the various backgrounds over which it passes.

PVC sheathed or other types of cables laid in roof spaces or under floors shall be set out neatly and systematically.

.03 Fittings

Switch plates, socket outlets and the like shall be fixed squarely and flush with wall surfaces.

Socket outlets shall be generally positioned 200mm above floor level or worktops, light switches 1400mm above floor level.

.04 Builders work

Cables under flooring shall be drawn through holes drilled at mid-depth of joists and not notched into the tops of joists.

Holes through existing or new floors, walls and ceilings shall be provided as required and made good whether or not specifically mentioned elsewhere.

Chases shall be provided in walls where required to hide wiring or conduit.

Suitable blockboard backboards shall be provided for meters and consumer units and the like.

Glazing

2.25 Materials

.01 *Clear sheet glass*

Clear sheet glass for glazing generally shall be ordinary quality (OQ) to BS 952.

.02 *Obscured glass*

Obscured glass for glazing to bathrooms and lavatories etc, shall be glass to BS 952.

.03 *Wired glass*

Wired glass shall be to BS 952. That for glazing to rooflights and the like shall be cast and for glazing to fire resisting doors and the like shall be Georgian wired float glass.

.04 *Putties*

Linseed oil putty shall be to BS 544.

Putty for glazing to metal shall be an approved proprietary brand.

2.26 Workmanship

.01 *Glazing*

Standards of workmanship generally shall comply to BS 6262.

.02 *Glazing with putty*

Glass is to be well back puttied and sprigged, front putties cut to clean lines and surplus back putties cut away.

.03 *Glazing with beads*

Glass is to be bedded in washleather strip and beads fixed firmly with brads or cups and screws as specified.

.04 *Wired and fluted glass*

Any glass with horizontal or vertical wires or patterns is to be cut so that the wires or patterns are parallel to the sash or other framing.

.05 *Glazing to old sashes*

Where cracked or broken glass is to be renewed the old putties shall be completely hacked out, care being taken in the case of wood sashes not to damage the rebates.

Painting and decorating

2.27 Materials

.01 Knotting

Knotting is to be in accordance with BS 1336.

.02 Stopping

Stopping for timber shall be composed of pure white lead and linseed oil putty (1.2) with a small proportion of cold size added or alternatively an approved proprietary stopping may be used.

.03 Generally

Paints are to be obtained from one manufacturer, approved and any instructions or recommendations strictly followed.

.04 Special paints or finishes

Specifier to list requirements.

.05 Surfaces to be emulsion painted

Unless otherwise defined in the schedule of works, all bare plaster surfaces, new or old, are to receive one mist coat and two full bodied coats of paint.

.06 Surfaces to be oil painted

Unless otherwise defined in the schedule of works, all new wood, metal or plaster surfaces are to be primed and painted two undercoats and one gloss finishing coat internally and primed and painted one undercoat and two gloss finishing coats externally and previously painted wood, metal or plaster surfaces are to be painted one undercoat and one gloss finishing coat.

.07 Surfaces to be varnished, lacquered or wax polished

Unless specifically stated to the contrary elsewhere two coats of varnish, lacquer and wax polish shall be applied. Exterior work which is to be lacquered shall have four coats.

2.28 Workmanship

.01 *Generally*

Standards of workmanship generally shall comply with BS 6150 and CP 5493.

.02 *Preparation of plaster surfaces*

All new plaster surfaces are to be thoroughly dried, brushed down, splashes of mortar, plaster etc removed and all holes, cracks and imperfections filled and made good before decorating.

Existing papered plastered surfaces are to be stripped by washing down, scraping and sand papering.

Existing painted plastered surfaces are to be washed down to remove dirt and grease and all loose paint removed by scraping and sand papering.

All nails, screws and plugs and defunct fixings are to be removed and all holes, cracks or imperfections filled and made good in existing plastered surfaces.

Existing water painted or distempered surfaces are to be sealed before paint of any type is applied or alternatively all traces of the old paint washed off. For the purposes of the number of coats of paint to be applied such surfaces shall be considered bare plaster.

.03 *Preparation of rendered surfaces externally*

New work which is to be painted is to be lightly brushed down and prepared in accordance with the paint manufacturer's instructions.

Existing surfaces are to be prepared by brushing to remove dust and loose paint. Any loose rendering is to be cut out and made good to match existing. Any defunct holderbats, screws, nails, plugs or other fixings are to be removed and the rendering made good. Defunct wiring and cable clips are to be removed.

.04 *Preparation of wood surfaces*

New woodwork to be painted shall be rubbed down with sand paper, knots covered with shellac knotting, surfaces stopped with suitable internal or external stopping as appropriate, rubbed down and cleaned off.

Open grain surfaces of plywood and the like shall be adequately filled.

New woodwork to be clear varnished, lacquered or polished shall be rubbed down with fine sand paper and any pin holes or small imperfections filled with matching colour filler.

.05 Preparation of existing wood surfaces

Existing painted woodwork shall be washed down to remove dirt and grease and rubbed down with pumice stone or waterproof abrasive paper and water to produce a smooth and level surface. Where existing paintwork is crazed, blistered or flaking then burning off must be carried out down to the bare wood, which should then be brought to a smooth and level surface by sand papering. All nails, cables, clips and other defunct fixings shall be removed and all holes and imperfections stopped and filled and rubbed down. Bare patches are to be primed and brought forward as necessary.

.06 Preparation of metal surfaces

All metal work to be painted is to be wire brushed to remove rust scale and cleaned down with white spirit to remove grease and dirt. Bare patches in old work are to be primed and brought forward as necessary.

Copper pipes shall be degreased with a solution of one part acetone to three parts benzole.

.07 Architectural ironmongery

Before the painting of wood or metal surfaces all ironmongery that is not to be painted is to be removed and cleaned and refixed when the paintwork is hard.

.08 Preparation between coats

The priming coat and each undercoat shall be well rubbed down with fine sand paper and stopped and touched up prior to the application of the succeeding coat of paint.

.09 Atmospheric conditions

No external painting is to be carried out in wet, foggy or frosty conditions. No painting externally or internally is to be carried out on surfaces which are damp.

.10 Paperhanging

The appropriate grade of adhesive must be used for various types and weights of wallpaper. Fungicidal adhesive must be used with vinyl coated papers.

Paper in any one room must be from the same batch to ensure proper matching of patterns and colours. Paper shall be hung butt jointed vertically and without horizontal joints, pattern joints must be properly aligned. The finished surface should be free from bubbles, tears and staining.

Drainage

2.29 Materials

.01 *Cement*

Cement shall be ordinary Portland Cement to BS 12.

.02 *Sand*

Sand for mortar shall be clean, sharp pit sand to BS 1199/1200 free of loam, dust or organic matter.

.03 *Aggregates*

Aggregates for concrete work shall be as described in Concrete Work.

.04 *Water*

Water shall be clean and free from all harmful matter.

.05 *Granular materials*

Granular materials for bedding and surrounds to drainpipes shall be a free running material to pass a 19mm sieve.

.06 *Drainpipes and fittings*

The choice of material is at the option of the builder providing it meets the approval of the local authority. Materials shall comply to any relevant British Standards.

.07 *Bricks*

Bricks for use in manholes small be local semi–engineering bricks suitable for use without rendering or local stock bricks if manholes are rendered internally.

.08 *Manhole covers*

Manhole covers and frames shall be iron to BS 497, grade C for pedestrian traffic, grade B for motor vehicular traffic. Double seal recessed pattern to be used internally, single seal pattern externally.

2.30 Workmanship

.01 Generally

Drainage systems shall be constructed in accordance with building regulations and by laws and as defined in the schedule of works and/or drawings. The builder shall allow for testing the completed installations to the satisfaction of the Building Inspector/Archt/SO.

Standards of workmanship generally shall be to BS 8301.

Drainpipe manufacturer's instructions and recommendations shall be followed.

.02 Drain runs

Drains shall be 100mm nominal minimum internal diameter, laid in straight lines to even falls of not more than 1:40.

Manholes shall be constructed at junctions and changes of direction.

.03 Manholes

Manholes shall be constructed of one brick thick walls on 150mm concrete base oversailed at top or provided with a concrete cover to receive a cast iron manhole cover and frame.

Internally the walls shall be fair faced engineering bricks or rendered in cement and sand (1:3) with proper cement rendered benching around main channel and branches.

Manhole covers shall be finished level with adjacent pavings, grass or earth.

.04 Connection to sewer

This work shall be carried out in conjunction with and to the requirements of the local authority and their charges paid by the builder.

Note: For vertical soil, waste and ventilating pipes and fittings, see Sanitary Plumbing 2.21.03.

Important

Since published standards are up-dated from time-to-time readers should check that they have access to the most recent editions of relevant standards, paying due regard to EEC directives.

Usage of sundry building terms

The following terms explain and amplify various terms and indicate the intention of section 3 "Work Parameters".

Removal includes:

dismantling/pulling down/taking down/taking out/taking up/stripping etc at the site of the works

getting from the site of the works to the outside of building by whatever means is necessary

disposal.

Disposal/Cart away – either of these terms include:

handling on site to store or to pick up point for loading

loading into skips or lorries

transporting away from site to yard, store or tip

payment of all tip charges.

Making out includes:

infilling to voids, openings, gaps and the like and matching materials and construction to existing.

Making good includes:

work as last described consequent on the carrying out of other work.

Form opening in brickwork or blockwork includes:

shoring up and needling as required

cutting the opening

designing, providing and inserting required beam or lintel and providing any calculations if required and obtaining building regulation approval providing and inserting cavity gutters and the like

forming new arches and the like in facework to match existing

quoining up jambs

sealing cavity of hollow walls, at jambs and cill and providing and inserting dampproof course

making good facework and features to match existing

forming new external sub–cills or sub–thresholds to match existing

making good the external plasterwork or other applied finishes including making out into reveals and providing metal angle beads to arrises where required.

and internally – at their respective locations "Making out/making good . . . in forming opening" in the various Trades (or under "Sundries") includes:

making good the internal plasterwork – as for external work last described

making good any matchboard, wallboard or other dry linings internally likewise

making out flooring into openings

carefully cutting out skirtings, picture rails, dado rails and the like and making good up to new opening including returning into reveals

removing debris.

Form opening in internal stud or framed partition includes:

shoring up as required

carefully cutting out and removing lath and plaster, plasterboard, matchboard or other facing material and afterwards making good around new opening

altering and adapting studding or framing as required to form opening sizing, providing and inserting required timber beams if the partition is loadbearing

and at their respective locations "Making out/making good . . . in forming opening" in the various Trades (or under "Sundries") include:

making out flooring into openings

carefully cutting out skirtings etc as described before

removing debris.

Block in/Blank off/Fill in opening in brickwork or blockwork includes:

carefully cutting out any flooring in opening and levelling and preparing for raising new work

cutting toothings for bonding in new work

filling the opening with brickwork or blockwork to match existing

making out facework including cutting out arches, cills or ornamentation around the opening and continuing any general facework pattern

wedging and pinning to existing soffit

providing and inserting matching damp proof course

making out any external plasterwork including continuing any existing patterns or labours and making good between new and old work so that after decoration or weathering the original opening cannot be discerned

and internally at their respective locations "Making out/making good . . . in blocking in opening" in the various Trades (or under Sundries) includes:

making out any internal plasterwork – as for external work last described

making out any matchboard, wallboard or other dry linings internally likewise

making out skirtings, picture rails, dado rails and the like.

Block in/Blank off/Fill in opening in stud or framed partition includes:

providing and inserting additional studding, plates and noggins or framing as required

and at their respective locations "Making out/making good . . . in blocking in opening" in the various Trades (or under "Sundries") includes:

making out plasterwork including making good between new and old work so that the original opening cannot be discerned after decoration

making out any matchboard, wall board or other type of linings likewise

making out timber skirtings, picture rails, dado rails and the like.

Adapt opening

This term is used to describe situations where existing openings are:

i. partly blocked in and reduced in size

ii. partly cut away and enlarged in size

iii. combination of (i) and (ii).

The meanings listed under form and block in opening shall apply as appropriate.

Remove chimney breast

This term includes:

pulling down brickwork flush to the nearest wall face

pulling down brickwork below floor level to allow the making out of floors

supporting any remaining breast at high level in a room by means of corbelling in traditional fashion including closing off any exposed flues, or in roof spaces suitable steel angle brackets may be used and flues left open when the flues are capped off

facing up exposed walls

and at their respective locations "Making out/making good . . . in removing chimney breast" in the various Trades (or under "Sundries") includes:

making out brick facings, plasterwork or other wall and ceiling finishes including cornices and other enrichments

filling in openings in floors and ceilings where breasts removed with timber floor or ceiling joists of scantlings to match existing properly framed in or supported on metal joist hangers as appropriate

making out floor boarding or solid floors and any applied finishes

making out skirtings, picture rails, dado rails and the like

removing debris.

Remove chimney stack

This term includes:

providing scaffolding fans and hoardings as required

removing pots

pulling down brickwork to below roof slopes

making out brick facings or rendering on adjacent walls where stacks were attached

filling in openings with timber rafters or flat roof joists of scantlings to match existing properly framed in

making out roof finishes, flashings or skirtings

removing debris and cleaning flues.

Remove brick or block wall or partition includes:

shoring up and needling as required

designing, providing and inserting required beam if the wall is load bearing and providing any calculations required and obtaining building regulation or by-law approval

*It is suggested that making out/making good areas of in-filling should be noted at their locations in schedules of work to draw the builders attention to them for sub-contracting purposes – as paragraph 3.33 "Works Parameters – Principles".

pulling down brickwork or blockwork to the extent indicated on the drawings or defined in the schedule of works

taking out any doors, lights, hatches and the like complete with frames or linings within the area of the wall to be removed

pulling down brickwork or blockwork below floor level to allow the making out of floors

facing up brickwork or blockwork where walls have been removed

quoining up free ends of walls left standing

and internally at their respective locations "Making out/making good . . . in removing wall" in the various Trades (or under "Sundries") includes:

making good brick facings, plasterwork or other wall and ceiling finishes including cornices and other enrichments

making out floor boarding or solid floors and any applied finishes

making out timber skirtings, picture rails and the like

removing debris.

Remove partition

This term in relation to stud or framed partition includes:

shoring up if required

sizing, providing and inserting required timber beam if the partition is loadbearing

taking off skirtings, picture rails and the like

stripping off lath plaster or other finishes and insulation quilts

taking out doors, borrowed lights, hatches and the like, frames, linings and architraves and the like within any area of partitioning to be removed

dismantling and taking down studding or framed work

and at their respective locations "Making out/making good . . . in removing partition" in the various Trades (or under "Sundries") includes:

making good plasterwork or other wall and ceiling finishes including cornices and other enrichments

making good or making out floor boarding and any applied finishes

making out timber skirtings, picture rails and the like

removing debris.

Block in/Blank off/Fill in fireplace opening

This term includes:

pulling down and getting out existing surrounds, mantles, grates, firebacks and hearths

sweeping the old flue

bricking up the opening with a half brick wall or blockwork properly toothed and bonded or studwork securely plugged and fixed

leaving an opening for and providing a fibrous plaster vent

making out brick facings, plasterwork or other wall finishes including making good between new and old work so that after decoration the original opening cannot be discerned

making out timber flooring to hearths including framing in any necessary joists or bearers

making out any applied finishes to boarded or solid floors

making out timber skirtings and the like

removing debris.

Seal flue

This term includes:

removing existing chimney pot (if any) and reinstating with (or providing new) ventilated and weathersealed pot including all necessary remaking of cement mortar flaunchings.

Repair roof covering

The term repair as applied to a tiled or slated roof includes any or all of the following operations as are necessary:

renew broken or missing tiles/states to match existing including nailing with composition nails or securing with copper tingles

re-wedge and repoint flashings and making out with new as required

re-make tile/slate verges or eaves including any bedding and pointing

renew defective or missing ridge or hip tiles

remove debris.

Renew roof covering

The term renew roof covering as applied to a tiled or slated roof includes:

lift and afterwards refix flashings, soakers, ridge, hip and valley coverings etc

strip existing roofing and battens, sort and set aside sound tiles/slates

renew battens and re-lay existing tiles/slates together with new tiles/slates as required all to match existing including sarking felt underlay whether previously provided or not, and including any special tiles/slates to eaves, verges, ridges and valleys.

re-wedge and repoint flashings

remove debris.

The term renew roof covering as applied to a sheet metal, felt or asphalt roof includes:

strip existing roofing

renovate sub-base as required

lift and afterwards refix flashings

renew roof covering to match existing

re-wedge and repoint flashings

remove debris.

Renew flashings and the like

The term renew flashings and the like as applied to pitched or flat roofs includes any or all of the following as may be applicable:

strip existing flashings, soakers, gutters, ridge and hip coverings

renew all work previously removed in material or similar quality and substance

re-wedge and repoint all new flashings

remove debris.

Ease and adjust

The term ease and adjust as applied to doors, cupboard doors, casement sashes and the like includes

rehanging on existing hinges

planing edges as necessary

oiling locks and hinges and leaving in working order.

Overhaul

The term overhaul applied to doors, cupboard doors, casement sashes and the like includes any or all of the following operations as are necessary:

cramp up loose tenon joints and wedge or re-wedge including gluing wedges

piecing in any damaged timber to door, frame and linings or architraves

rehanging on existing hinges or renewing hinges if required

plane edges

plane off protruding tenons

refix ironmongery and locks or renew if required

oil locks and hinges

renew glass where cracked or broken

renew putties where loose, missing or defective.

The term overhaul applied to double hung sash windows means any or all of the following operations as are necessary:

renew sash lines, pulleys and weights

renew parting beads

renew inner beads

renew missing or defective sash fasteners and pulls

ease and adjust sashes

piece-in any other defective woodwork

renew glass where cracked or broken

renew putties where loose, missing or defective

Remove staircase

Remove staircase includes:

dismantle and take down staircase including spandril panelling and balustrading, apron linings and all associated work

frame in floor joists to fill, alter or adapt stairwell opening as required by the drawings or schedule of works

make out floor boarding to suit

make out skirtings, picture rails and the like to suit

make out plastered ceilings

make out any unplastered walls exposed by the removal

making good plasterwork where fixings, bearers etc are removed

making good between new and old plasterwork so that after decoration the old work is not discernable

remove debris.

Strip existing installation

The term strip existing installation in relation to electrical installation includes:

disconnecting at mains and making safe

disconnecting and taking out all existing conduit, wiring and fittings (except where conduit is to be re-used)

remove existing fuseboards

remove defunct clips, fixings and the like

making good walls, floors, ceilings as required

removing debris.

Strip existing installations in relation to plumbing and engineering installations shall allow for:

turning off incoming supplies

disconnecting and taking out all existing appliances, fittings and pipework

removing defunct pipeclips, fixings and the like

making good walls, floors, ceilings as required

removing debris.

Section Three

Schedules of Work – Presentation and Order

Introduction

Specifications can be reduced in size and content by sub-dividing a building into composite parts and defining their content (or parameters).

This section deals with:

(i) the concept in principle;
(ii) general requirements associated with any work, such as regulations, obligations, provisions of facilities such as scaffolding, disposal of rubbish and so on;
(iii) an order for scheduling work that additionally provides a procedure for building surveys and a checklist;
(iv) the parts and ancillary parts comprising composite items: for example, a bath including its taps, trap, waste, panels and plumbing connections.

Work Parameters – Principles

3.01 This section sets out the PRINCIPLES determining the sub-division of a building into sections and sub-sections; the order for schedules of work – the widest possible application of "spot pricing" by defining the parameters of composite items of work and what items may be *deemed to include* for this purpose. The order can also apply to survey.

3.02 In order to achieve standardisation, the need is to obtain the broadest possible adoption of such sub-divisions and ordering of work. To this end, the order of EXTERNAL and INTERNAL works is a broad sub-division already frequently followed and when STRUCTURAL work arises this in turn, logically precedes other work. The basis for the main ordering of work may thus be simply stated as follows:

EXTERNAL WORK

predominantly STRUCTURAL

under LOCATIONAL headings depending on scope of work

INTERNAL WORK

predominantly LOCATIONAL

but

STRUCTURAL work takes precedence internally – for example when adapting to a new layout as in "conversions".

3.03 This order also coincides with a sub-division to parts which are "COMMON" (being mainly structural) and others which are "UNIQUE" – as shown diagrammatically in Fig 2, page 138. Both these sub-divide to clearly identifiable building parts, or elements, producing recommended STANDARD SUB-DIVISIONS. These are detailed in this document and are recommended for both survey and schedules of work, but if circumstances vary the order of survey, change to the prescribed order can be made in the schedule. Also, if the size of a job warrants, further sub-division can be made.

3.04 Notwithstanding the policy to composite items, it is found that a division between External and Internal works is most positive at the interface between the external structural fabric and the internal finish; so that, generally, structural work to such a wall is External and the work to its inner face like plaster, is Internal work.

3.05 Damp Proof Courses by insertion, injection or other method occurring in external and/or internal walls, being items relevant to structure, are dealt with under a separate heading in Internal Works – because incidental work to their installation is generally internal. Damp Proof Membranes (unless part of a tanking job to a basement – under a separate heading) are taken as part of the composite floor item located on the lower-most floor. The incidental work to the separate compartments (or rooms) arising from damp course work, other than that disturbed for purposes of access to do the work, is described at the various locations.

3.06 Internal main dividing walls and floors, which result in the formation of compartments producing locationally self-contained work sections, take precedence. Partitions are treated as a sub-division of compartments to create rooms subsequently.

3.07 Thus, location of work internally is usually by the sundry spaces a building is divided into so that where a larger space is divided to form a self-contained unit, or for example, where a large room is divided into a bedroom, bathroom and lobby, the larger space is the foremost locational heading and smaller spaces subsequent headings.

3.08 Work internally, generally proceeds by compartments from top downwards with the Staircase as an entity – generally last.

3.09 Broadly, the order is therefore compartments, followed by partitions, followed by rooms; the order in each room following a suitable Trade sequence, such as:

Fireplace
Plaster and screeds
Flooring, coverings, insulation, etc
Windows (and frames)
Doors (and frames)
Borrowed lights etc (and frames)
Cupboards
Skirtings, dadoes, picture rails and the like
Sinks, baths and other sanitary fittings
Services (hot and cold water, heating, gas and electrics)
Sundries
Repairs
Decorations to ceilings, walls, paintwork, etc.
Note: Staircases should follow a similar order with the stairs included in the appropriate Trade sequence.

3.10 If work is extensive, it is preferable to insert in the schedule of works the sub-heads above for ease of identification. Also, if work to windows, doors, services and the like is sufficiently extensive, it should be referenced to separately tabulated schedules for each, annexed to the schedule of works.

3.11 Sundries: Inevitably, there will be miscellaneous items of work (other than repairs) that will need to be carried out to complete a job and additionally, to meet the full requirements of the client. Each item should be scheduled under the sub heading "Sundries".

3.12 Repairs: This term is widely used to mean the restoration of any building parts to sound and stable condition generally involving the marrying-in of new to old work. Thus, it may range from a major piece of new work to sundry minor items. Distinctions, therefore, are made to limit its application, as follows:

NEW WORK is regarded as the provision or installation of any part not previously existing, for example: a built-on extension; a central heating installation or the first-time provision of a bath

RENEWAL is considered to apply to the rebuilding or reinstatement of a major part or element, for example: the rebuilding of an entire facade, the recovering of a roof or the renewal of a "rotted" wooden floor.
Note: There may be alternatives to the term "Renewal" – see Glossary of Terms later.

REPAIR is applied to more minor items generally relating to part of an element, for example: the "stitching" of a crack in brickwork, the renewal of a sash to a window, the easing of a door or renewal of items of ironmongery.

3.13 Some works, such as Maintenance, may be predominantly Repair, when further sub-heads may be used to locate the items.

3.14 Drainage: Works around the Buildings, etc. should follow an order of work under sub-headings, for example:
Drainage; Outbuildings (including garages); Pavings; Garden Works; Boundary Walls; Fences; etc.

3.15 Generally: The extent of all works must be established and scheduled as fully as it is possible to determine. Dimensions should be given when necessary to identify the extent of work and the sizing of parts. Any faults should be fully inspected and diagnosed as to cause and remedies determined at the outset so as to firmly schedule the work. Where the builder prepares the schedule and is to carry out the work, he is deemed to have included for everything possible to execute and complete the works. In all cases, the builder is deemed to have visited the site and inspected the work before submitting his tender.

3.16 Where probable work is covered up, such as in situations underground, the best judge should be made of requirements and stated as a basis of any final adjustment of costs; for example, the depth of foundations.

3.17 Demolition and Stripping Out: It should be incumbent on the client to specify anything he has a wish to retain and such items should be scheduled, otherwise the ensuing rubbish or debris shall be removed by the builder.

3.18 Some existing materials may be reusable. It should be stated wherever it is proposed to incorporate these in new or reinstated work.

3.19 Tendering: The object should be to produce proposals for work as realistic and as firm as possible and tenders should ordinarily be FIRM. They should be based on an itemised build-up of the tender and this should be made available before work is put in hand, as a basis for accounting for any change arising in the subsequent scope of the work.

3.20 Ordering of Sections in the Schedule of Works: The order of sections summarised on pages 143–153 should be followed as standard.

Section 3: House Improvements and Conversions

DIAGRAMMATIC REPRESENTATION of the PROBLEM of:

1 Analysing to a Rational Classification of Works Headings;
2 Analysing Costs to a Rational Classification of Works Headings;
3 Assessing Technical Worthiness of Proposals;
4 Assessing Design Worthiness of Proposals;
5 Presentational Order of Inter-related Works;
6 Determining Extent of Work *(including structural content in varying degrees)* and Unit Size of Dwelling and of Measurement.

Bungalow
House
Multi-occupied
• Roof level
• Facade level
• Foundation level
M/ette
Flat
Flat
Flat
Flat
Flat
Flat
Flat
Flat

UNIQUE PARTS:
(serving individual units):
*Bath
*Basin
*Sink
*Hot and Cold Water Supplies
*WCs (Internal/External)
*Electric Light and Power
*Heating-by Open Fire or Appliance; or Central
*Fuel Storage
*Refuse Storage and Disposal

(and possibly with related common services)

*Kitchen Fittings and Food etc. Storage
*Windows (and vents) – for *Natural Light and Ventilation*
External Doors

Ground Floors
Staircases
Upper Floors
Partitions

Internal Doors
Finishes – Ceiling; Walls
Decorations

COMMON PARTS:
(serving several units)
at ROOF level:
Roof, Gutters and Downpipes; Chimney Stacks; Roof Insulation

at FACADE level:–
Main Walls *(to be structurally stable*)*
Common Staircases *and their Windows and External Doors*
Balconies
Lifts and Lift Towers
Fire Escapes
Insulation-against Fire; Heat Loss & Sound

Service Stacks – Water; Electric; Gas; Soil; Refuse; etc.

Dampproofing; Damp Courses *and *associated remedial works*
Paintwork

at FOUNDATION level:–
Foundations *(to be structurally stable*)*
Drainage*
Externals

IMPROVEMENTS	may be to varying types of house occurring singly at random, OR in groups OR estates OR be to varying types of flat or maisonette at random OR in stacks of varying height OR comprise one or more stacks in blocks OR estates.
CONVERSIONS	may occur in any of the above and be any combination of contiguous types
ARRANGEMENT	may involve: (i) works to eradicate 'unfit' features – eg trip steps; dark winding stairs; etc. (ii) built-on Extensions or Adaptations within existing fabric. (iii) lifts and lift towers and formation of access. (iv) fire escapes and access.
REPAIRS	any unique or common part may incur repair in varying degrees, or renewal and may include treatment of woodworm or dry rot.
COSTS	are influenced by size and type of scheme and commensurate building firm; market conditions etc.
KEY:	Housing Act '80 'Basics' (ie bath; basin; sink; inside WC; hot & cold supplies) and '10 points' are starred thus*.

Work Parameters – General Requirements

3.21 The concept of building elements is already long established. For measurement in respect of superstructure, these are:

3.22	Vertical Elements:	between the horizontal plane at damp course level and roof plate level.
	Horizontal Elements:	on the horizontal plane within main enclosing walls. In regard to roof construction for example, this includes wall plates, rafters, ceiling joists, etc, firewalls and parapets.

3.23 The several commonly adopted main elements for cost analyses are followed but with appropriate modification for these WORK PARAMETERS. As stated in the Introduction, they represent composite building parts; for example, sanitary fittings complete with trap, taps and short service connections (or "tails"); elements of the structure; etc.

3.24 Parts of a composite item, like ironmongery to a window, usually arise as items of repair and come under a heading "Repairs"; or if new, come under a heading "Sundries".

3.25 Pricing should be itemised by the builder. He should allow for the requirements of regulations and apply his expertise in regard to methods of fixing, sizing of pipes, matching of bricks, etc (unless otherwise specified).

3.26 Pricing should also allow for all costs of labour, materials, setting out, administration, supervision, office overheads, profit, site requirements, scaffolding, ladders, tackle and other plant, attendance of one trade upon another, cutting away and making good, protection and screens, clearing of rubbish, cleaning up and for everything required to carry out and complete all items of work.

3.27 Where there is salvage value for anything to be removed, this is to be deemed allowed for in the price. Items to be salvaged and retained by the client should be defined.

3.28 Renewal is generally deemed to mean reinstatement as existing and to require new materials; departures from this which include the reuse of existing materials, fittings, etc and/or changes in type, style and design will need to be defined or shown on drawings.

3.29 Items to be reused are to be carefully handled and stored pending reuse – see Work Parameters – Principles.

3.30 Current regulation regarding thickness, etc will apply – invariably where any element is changed or substantially renewed.

3.31 Toothing, bonding, matching and marrying to existing work in small areas; hacking or leaving key for finishes; the making good of holes, chases and other work cut away by the builder in carrying out and completing an item of work should all be deemed included but not work due to defects requiring repair, such as "live" plaster, which should be separately scheduled under "Repairs".

3.32 All new work, including new material in repairs, should be primed before fixing. Timbers to be treated with preservative should be separately defined.

3.33 As stated in "Principles", internal finishings are separated at their interface with structure, so that internal plastering is kept separate from external walls. This is not only more definitive but plastering, frequently sub-contracted, is thereby more readily identified and grouped as a trade for tendering purposes. Likewise, ceiling finishes are separated from floor or roof construction.

3.34 Where work of a specialist kind is to be carried out, a specification and firm price should be obtained and included by the builder in his tender with the addition of cash discounts allowable to him. A copy of the specification should be supplied to the builder to ensure he provides facilities, attendance and supplies not included in the specialist quotation. Where the builder is preparing the schedule of works, he should include similarly for any specialists work, facilities, etc.

3.35 Heating and hot water, gas and electrical installations will frequently be covered by separate and complete specifications. These should be appended to the schedule of works and a note made where the points are located and cross referenced to the respective Service specifications attached.

3.36 In "minor works" the builder should be responsible to ensure the full execution and completion of the works and payment of all outgoings incurred. Where the client directly sub-lets any part of the work, the separation of responsibility and charges to be paid should be clarified at the outset and defined in the schedule.

3.37 The responsibility of the builder for his work includes safety and stability during its execution. He should therefore allow for any shoring, strutting, support and screens in forming openings, in works of underpinning or any rebuilding and the like. The same responsibilities also apply to him where he himself employs a specialist or sub-contractor for any part of the work.

3.38 Responsibility for insurances is defined in forms of contract and will depend on the form used. In the event of a client employing a builder using none of these forms, arrangements for insurance should be settled at the outset and defined in the schedule; for example, by following the insurance requirements of the JCT Forms where no architect/supervising officer is employed; or by resorting to the BEC Guarantee Scheme or FMB Warranty.

3.39 The builder will be responsible for complying with statutory notices and for paying any fees applying to the works but the client should be responsible for by-law and planning approvals and applying for grant where this is sought towards improvement and/or repair etc. also, for entreating with party owners, etc.

3.40 VAT: Tenders (or Estimates) should exclude VAT. Accounts when rendered should separately record VAT and conditions relating to VAT should, for example, follow the JCT Contract Forms.

Standard Order for Scheduling Work

NOTES:

1. *The order need not be rigidly followed on survey if separate field sheets are headed with the work sections shown and shuffled into the above order for the final schedule of works.*

2. *The use of paper with feint lined columns for recording similar items of work at different locations, or for allocating costs under different account headings, etc is described on page 5 and shown in Fig. 1, page 64.*

Externally

3.41 UNDERPINNING: EXTERNAL WALLS

3.42 STRENGTHENING: EXTERNAL STRUCTURAL PARTS

3.43 DEMOLITION: MASS REMOVAL OR RANK PULLING DOWN

3.44 PULLING DOWN/TAKING OUT/STRIPPING/ETC: PIECEMEAL REMOVAL

3.45 EXTERNAL WALLS:

.01

Renewal: Rebuilding of major parts or elements. May include foundations.

.02

Isolated Openings: Formation, enlarging, reducing or blocking up.

.03

Repairs: Restoration of *minor parts* of major parts or elements.

.04

Insulation:

3.46 ROOFS:

.01

Chimney Stacks – Taking down:

.02

Chimney Stacks – Renewal:

.03

Chimney Stacks – New:

.04

Roof Coverings – Renewal:

0.5

Roof Construction – General:

.06

Roof Construction – Renewal:

.07

Eaves, Gutters and Downpipes – Renewal:

.08

Repairs: Grouped – relating to above sub-headings to "Roofs"

.09

Insulation

3.47 NEW STRUCTURES:

.01

Extensions: There may be one or more: Each should be dealt with under a heading in the order below of composite parts or elements:

Foundations
Ground Floor
Upper Floors
External Walls
Roofs
Eaves; Gutters and Down pipes
Insulation

3.48 SUNDRIES:

These may be a miscellany of items such as canopies, porches, balustrades; smaller items like chimney cowls or more substantial items such as may occur in buildings of three or more storeys like lifts and lift towers, refuse chutes, fire escapes, etc. Where appropriate the works should be under sub-headings.

3.49 PAINTWORK:

Internally

Note: The sub-division of internal finishes and external fabric is at their interface.

3.50 UNDERPINNING: INTERNAL WALLS

3.51 STRENGTHENING: INTERNAL STRUCTURAL PARTS

3.52 PULLING DOWN/STRIPPING OUT/TAKING OUT/ETC: PIECEMEAL REMOVAL

3.53 INTERNAL WALLS:

.01

Removal:

.02

Renewal: Rebuilding (or repositioning) major parts or elements. May include foundations.

.03

Isolated Openings: Formation, enlarging, reducing or blocking up.

.04

Repairs: Restoration of *minor parts* of major parts or elements.

.05

New Walls:

3.54 FLOORS:

.01

Removal:

.02

Renewal:

.03

Isolated Openings:

.04

Repairs:

.05

New:

.06

Insulation:

3.55 CHIMNEY BREASTS:

.01

Entire Removal:

.02

Renewal:

.03

New:

.04

Repairs:

3.56 DAMP PROOF COURSES IN EXISTING WALLS: EXTERNAL AND INTERNAL WALLS – USUALLY A COMBINED OPERATION

3.57 COMPARTMENTS: USUALLY STRUCTURALLY SELF-CONTAINED SPACE THAT IS FREQUENTLY SUB-DIVIDED BY PARTITIONS INTO ROOMS.

Note: After structural sub-division is complete the order for compartments is generally top storey downward.

3.58 PARTITIONS: USUALLY MORE CONVENIENTLY GROUPED UNDER A HEADING:

.01

Partitions – General:

.02

Partitions – Renewal:

.03

Partitions – Isolated Openings:

.04

Partitions – Repairs:

.05

Partitions – New:

.06

Partitions – Insulation:

3.59 ROOMS: THE FOLLOWING WORKS SHOULD BE SCHEDULED UNDER EACH ROOM TITLE AND STOREY BY STOREY FROM TOP DOWNWARD WITH THE COMPARTMENT IN WHICH LOCATED GIVEN AS A HEADING:

.01

Fireplaces – General:

.02

Fireplaces – Blocking-in:

.03

Fireplaces – Renewal:

.04

Fireplaces – New:

.05

Wall Plaster, Etc – General:

.06

Wall Plaster, Etc – Renewal:

.07

Wall Plaster, Etc – New:

.08

Ceiling Plaster, Etc – Renewal:

.09

Ceiling Plaster, Etc – New:

.10

Screeds – Renewal:

.11

Screeds – New:

.12

Insulation:

.13

Flooring – General:

.14

Flooring – Covering – Renewal:

.15

Flooring – Covering – New:

.16

Flooring – Insulation:

.17

Windows (and Frames) – Renew:

.18

Windows (and Frames) – New:

.19

Doors (and Frames) – Renew:

.20

Doors (and Frames) – New:

.21

Borrowed Lights (and Frames) – Renew:

.22

Borrowed Lights (and Frames) – New:

.23

Cupboards, Kitchen Fittings, Etc – Renewal:

.24

Cupboards, Kitchen Fittings, Etc – New:

.25

Skirtings, Dadoes, Picture Rails, Trim, Etc – Renewals:

.26

Skirtings, Dadoes, Picture Rails, Trim, Etc – New:

.27

Stairs – General:

.28

Stairs – Renewal:

.29

Stairs – New:

.30

Sanitary Fittings – General

.31

Sanitary Fittings – Renewal:

.32

Sanitary Fittings – New:

3.60 MISCELLANEOUS SUNDRIES AND REPAIRS:

.01

Sundries: Grouped in Trade sequence covering miscellaneous items *other than repairs.*

.02

Repairs: Grouped in Trade sequence covering above sub-headings to "Rooms"

3.61 SERVICES: MAY HAVE MAINS COMMON TO SEVERAL COMPARTMENTS AND SUBJECT TO SEPARATE CONTRACTS, AS SET OUT BELOW:

3.62 COLD WATER SUPPLY:

.01

Cold Water Supply – Renewal:

.02

Cold Water Supply – New:

3.63 HOT AND COLD WATER INSTALLATION:

.01

Hot and Cold Water Installation – General:

.02

Hot and Cold Water Installation – Renewal:

.03

Hot and Cold Water Installation – New:

3.64 CENTRAL HEATING INSTALLATION:

.01

Central Heating Installation – General:

.02

Central Heating Installation – Renewal:

.03

Central Heating Installation – New:

3.65 GAS INSTALLATION:

.01

Gas Installation – General:

.02

Gas Installation – Renewal:

.03

Gas Installation – New:

3.66 ELECTRICAL INSTALLATION

.01

Electrical Installation – General:

.02

Electrical Installation – Renewal:

.03

Electrical Installation – New:

3.67 DECORATIONS:

Drainage, etc

3.68 DRAINAGE, OUTBUILDINGS, WORKS AROUND THE BUILDINGS AND TO BOUNDARIES:

.01

Drainage – General:

.02

Drainage – Renewal:

.03

Drainage – New:

.04

Outbuildings, Etc:

Standard definitions – externally

3.41 Underpinning

The extent should be ascertained and depth should be determined from trial holes.

The method of underpinning for "minor works" should be within the expertise of the builder who should be responsible for maintaining support and safety of the structure at all times and for satisfying building regulations and by laws.

If the work is beyond his scope, a specialist may be employed when a specification and firm quotation should be obtained and any separation of the respective work parameters established and defined.

Whether by a builder or specialist, it should be ensured that the incidental cutting away, setting-aside and reinstatement of items to gain access for the work is additionally covered by them.

The SCHEDULE should define the walls to be underpinned, the extent and the depth down to a firm sub-stratum and be *deemed to include* all work to gain access to the existing foundation or footings, keeping excavations free of water, cutting-off projecting footings, all further excavation, concrete, brickwork and wedging-up in underpinning, in alternate bays of short length, the backfill and disposal of spoil, etc and reinstatement of works disturbed.

Note: 1. The depth of foundations should be stated – see "Works Parameters – Principles".

2. Repairs arising from the cause to underpin, such as the "stitching" of cracks in brickwork, or the wedging of open arch joints, should be included under "Repairs" to external walls and/or elsewhere.

3.42 Strengthening

Such work will be varied.

If the work is beyond the builder's scope, a specialist may be employed when a specification and firm quotation should be obtained and a separation of the respective work parameters established and defined.

Whether by builder or specialist, it should be ensured that the incidental cutting away, setting-aside and reinstatement of items to gain access for the work is additionally covered by them.

The SCHEDULE should describe the work and be *deemed to include* all work to gain access and reinstatement of works disturbed.

Note: Repairs arising from the cause to strengthen should be included under "Repairs to External Walls" and/or elsewhere (as described for underpinning above).

3.43 Demolition

Demolition is considered to mean the mass removal of a building or part of a building by rank pulling down and removal of debris.

"Minor Works" demolition should be within the expertise of the builder who should be responsible for maintaining support and safety of standing structures at all times and for satisfying building regulations and by-laws.

If the work is beyond the builder's scope a specialist may be employed when a specification and firm quotation should be obtained and any separation of the respective work parameters established and defined.

The SCHEDULE should define the section of the building, the type of construction, its extent and approximate dimensions and be *deemed to include* cutting-off of supplies, dismantling from any parts to be retained, all cutting away, demolition of parts, getting out and disposal.

The manner of disposal after getting out should be described stating whether on or off site and how any items are to be retained.

3.44 Pulling Down/Taking Out/Stripping/etc

Pulling down, etc is considered to refer to the piecemeal removal of parts to be generally disposed of, (*as distinct from* parts to be renewed, ie pulled down etc and subsequently renewed and scheduled as composite items under the heading "Renewal").

The SCHEDULE should define the parts and type of construction, the extent and location and approximate dimensions and be *deemed to include* cutting off of supplies, dismantling from any parts to be retained, all cutting away, getting out and disposal, cropping back and finishing fair at exposed faces.

The manner of disposal after getting out should be described stating whether on or off site and how any items are to be retained.

The insertion of beams and their encasing where load bearing parts are removed should be separately defined.

Note: For Openings – see "Isolated Openings in External Walls" below.

3.45 External Walls

.01

Renewals: Refers generally to major elements eg facades of buildings.

The SCHEDULE should give the location and extent of the work and depth of foundations to a firm sub-stratum and be deemed to include taking down the prescribed parts, grubbing-up foundations, the provision of new work including excavation and keeping free of water, backfill, disposal of spoil and temporary supports, concrete and footings, damp courses (built-in), brickwork and/or stonework, etc, lintels, cills and work to openings, external finishes and features, building-in vents, ends or edges of abutting floors, etc, the setting-in position and pointing of any window and door frames.

Note: The depth of foundations should be stated – see "Work Parameters – Principles"
For New Works – see "Extensions"
For Gutters and Drainpipes – see "Roofs"
For Internal Finishes – see "Internally"
For Windows and Doors – see "Internally"

.02

Isolated Openings in External Walls: Refers to the formation, enlarging, reducing or the blocking-up of isolated opening in existing walls. The SCHEDULE should give location and size and be *deemed to include* cutting, adapting, forming or filling the opening; lintels, cills, facings and finishes externally and the setting in position and pointing of any window and door frames.

Note: For Internal Finishes – see "Internally"
For Windows and Doors – see "Internally"

.03

Repairs: These will generally involve the patching and matching of portions of walling and external surfaces like pointing, stuccowork and rendering; the stitching of cracks in walls; sundry works to steps, ironwork, etc. Each item should be SCHEDULED giving location and extent and be *deemed to include* pulling down and cutting away, hacking off, preparation, dubbing out and matching and marrying of new to existing facings, facework, mouldings and ornament.

.04

Insulation: The requirements of thermal insulation, grade and construction should be additionally defined.

3.46 Roofs

.01

Chimney Stacks – Taking down: The SCHEDULE should state the location, number of flues, approximate size and height and be *deemed to include* for taking down, capping off and providing vents to flues.

Note: For the removal or the renewal of chimney stacks – as a consequence of the removal or renewal of chimney breasts – see "Internally".

.02

Chimney Stacks – Renewal: The SCHEDULE should state the location, number of flues, approximate size and height and be *deemed to include* for taking down, rebuilding, parging of flues, damp courses (built-in), chimney pots, and flaunching, chimney gutters and flashings to stack, rough rendering inside the roof, the making good of the roof construction, covering and finishings around the stack, the redressing of plumbers metalwork and the clearing of the flues, served by the stack, of debris.

New plumbers metalwork should be separately defined.

.03

Chimney Stacks – New: The SCHEDULE should state location, etc as for "Renewal" and be *deemed to include* as there defined, excluding taking down but including cutting back, to join up with existing and all preparation.

.04

Roof Coverings – Renewal: The SCHEDULE should locate and define the extent of the work and the type of covering and be *deemed to include* stripping and renewal including underlinings, fixing battens, finishings at edges, abutments and valleys, hip and ridge cappings, fillets and skirtings and the redressing of plumbers metalwork.

New plumbers metalwork should be separately defined.

Note: For New Coverings – see "Extensions".

.05

Roof Construction – General: The inevitable stripping and recovering of a roof to carry out work to the construction should be dealt with under "Roof Coverings" as defined above.

If there is more than one roof, or work is to be done in sections, or is to form part only of a roof, each should be located, defined in extent and described.

.06

Roof Construction – Renewal: The SCHEDULE should define the type of roof and extent of renewal (but for firewalls and parapets, see below) and be *deemed to include* dismantling and reconstruction of all framework, joists, or slab, underlinings, eaves, plates and eaves filling, ceiling construction (but not ceiling finishes – see note below) and screeds.

Firewalls and parapets should be separately scheduled.

Note: For Temporary roofs and coverings for protection
– see "Work Parameters – General".
For ceilings – see "Internally".
For Strengthening (which generally retains existing construction)
– see "Repairs" to Roofs.
For New Construction – see "Extensions".

.07

Eaves; Gutters and Downpipes – Renewal: The SCHEDULE should state if to be removed in the form of the existing eaves, otherwise to define the form of construction giving size of parts, type and quality including gutters, downpipes and fittings and be *deemed to include* for stripping and renewal of fascias, soffites, gutters, downpipes and fittings.

Note: Individual parts will invariably arise as repairs and be dealt with under that heading.
For New Construction – see "Extensions".

.08

Repairs: These will generally involve patching and matching portions of coverings, construction, gutters, etc. Each item should be SCHEDULED, describing the location and extent and be *deemed to include* pulling down and cutting away, preparation, making out and matching new with existing.

Note: Strengthening is dealt with generally as described under "Strengthening".

.09

Insulation: The requirements of thermal insulation, grade and construction should be additionally defined.

3.47 New Structures

.01

Extensions: A drawing would normally be provided but the SCHEDULE should give a brief description of location, size and height, number of storeys and where not built of materials to match the existing, should define their type, quality and finish under their respective composite element headings set out below under Foundations, External Walls, etc.

Extensions, if non-traditional, should be fully described.

Any demolition will already have been dealt with under that heading.

The composition of the various elements of the new construction should be *deemed to include* as follows:

Foundations: Excavation in any material, planking and strutting to trenches, keeping excavations free of water, backfill and disposal of spoil, concrete, brickwork and/or stonework, etc including footings and facework up to and including damp course (built-in), concrete fill to cavity and any under floor ventilation.

Note: For Depth of Foundations – See "Work Parameters – Principles".

Ground Floor: Excavation in any material and disposal of hardcore and concrete beds.
If the construction is "solid", it should include damp proof membrane, screed and floor finish; if "suspended", the sleeper walls, damp course, sub-floor ventilation, joists and boarding. *It should be stated* if solid or suspended, and give the finished level in relation to existing floor level.

Note: For Skirtings – see "Internally"
For Floor Coverings – see "Internally"

Upper Floors: Joists and boarding/slab and screed; stating which form of construction and defining the formation and finish to openings.

Note: For Skirtings – see "Internally"
For Floor Coverings – see "Internally"
For Ceilings – see "Internally"

External Walls: Brickwork, stonework, blockwork and cavities, lintels, cills and work to openings, external finishes and features, building-in air bricks, ends and edges of abutting floors, etc and the setting-in position and pointing of window and door frames.

Note: For Internal Finishings – see "Internally"
For Windows and Doors – see "Internally"
For Skirtings, etc – see "Internally"

Roofs: Coverings, underlinings, battens, framework, joists or slab, eaves plates and eaves filling, screeds, ceiling construction (but not ceiling finish – see below), finishings at edges, abutments, valleys, hip and ridge cappings and all plumbers metalwork.

Firewalls and parapets should be separately defined.

Note: For Ceiling Finish – see "Internally"

Eaves, Gutters and Downpipes: Eaves construction, bracketting, fascias and soffites, gutters, downpipes and fittings.

Insulation: The requirements of thermal insulation, grade and construction should be additionally defined.

Note: For Partitions, Fitting-out, Internal Finishes and Services – see "Internally"
For Drainage, Works around the building and out buildings – see "Drainage" etc.

3.48 Sundries

This should include items generally related to the fabric of the building, not included above; for example, steps, window balconies, balustrades, canopies, door cases, porches, minor items like chimney cowls, etc.

Note: For toothing, bondings, etc to existing buildings see "Work Parameters – General".

In improvements or conversions to certain buildings such as blocks of three of more storeys, the following external structures may arise and be the subject of sub-headings:

Balconies
Common staircases
Lifts and lift towers
Service stacks such as for refuse
Fire escapes

They should be fully defined under sub-headings.

3.49 Paintwork

This will invariably be necessary on both new and old surfaces of the various materials to be painted.

The extent, type and quality of finishes (including undercoats) should be defined and be *deemed to include* the bringing forward of old painted surfaces, cleaning down, knotting, stopping and rubbing down for the finishes scheduled.

Burning off and subsequent preparatory work should be separately defined.

Note: All new work to be primed before fixing see "Work Parameters – General".

Standard definitions – internally

3.50 Underpinning – generally as for "EXTERNALLY"

3.51 Strengthening – generally as for "EXTERNALLY"

3.52 Pulling Down/Stripping out/Taking Out/etc

The terms are considered to refer to the piecemeal removal of parts to be generally disposed of; for example, a portion of a wall, floor or partition, window, cupboard or stair, electrical or other installation, etc *as distinct from* parts to be renewed, ie pulled down, etc and renewed and scheduled as composite items under the heading "Renewal".

For locational purposes, it is preferable to include these in the room, or compartment, where they exist.

The work should be within the expertise of the builder who, wherever structural stability is involved, should be responsible for maintaining support and safety of standing structures at all times and for satisfying building regulations and by-laws.

If the work is beyond the builder's scope, a specialist may be employed when a firm quotation should be obtained and any separation of the respective work parameters established and defined.

The SCHEDULE should define the parts of the building, the types of construction, extent and approximate dimensions and be *deemed to include* cutting-off of supplies, dismantling from any parts to be retained, all cutting away, getting out and disposal and cropping back fair at exposed faces.

The manner of disposal after getting out should be described, stating whether on or off site and how any items are to be retained.

The insertion of beams and their encasing – where load bearing parts are to be removed – should be scheduled.

Note: For Isolated Openings – see headings below
For Finishes and/or Sundries – see headings below

3.53 Internal Walls

REFERS GENERALLY TO MAIN STRUCTURAL WALLS AND DIVISION WALLS, POSSIBLY EXTENDING THROUGH ONE OR MORE STOREYS AND/OR BETWEEN MAIN COMPARTMENTS – AS DISTINCT FROM PARTITIONS – GENERALLY BETWEEN ROOMS.

.01

Removals: *see "Pulling Down, Etc", p. 166.*

.02

Renewals: The SCHEDULE should give the location and extent of work and depth of foundations to a firm stratum and be *deemed to include* taking down the prescribed parts, the grubbing up of foundations, the provision of new work including excavation and keeping free of water, backfill, disposal of spoil, strutting to trenches, concrete and footings, damp courses (built-in), brickwork, blockwork, etc, lintels and work to openings, building-in ends or edges of abutting floors, etc, the setting-in position and pointing of frames to borrowed lights, doors, etc.

Note: As main walls may pass through more than one storey, their finishes are dealt with in the respective rooms and compartments later.

For Depth of Foundations – see "Work Parameters – Principles"
For Finishes, Etc – see headings below
For Borrowed Lights, Doors, Etc – see headings below

.03

Isolated Openings in Existing Walls: Refers to the formation, enlarging, reducing or the blocking-up of isolated openings in existing structural or division walls.

The SCHEDULE should give location and size and be *deemed to include* cutting, adapting, forming or filling the opening, lintels, finishes and features to head, reveals and thresholds and setting-in position and pointing of frames to borrowed lights, doors, etc and the making-out of finishes where openings are filled, or partly filled-in.

.04

Repairs: These, in respect of main structural or division walls, will generally involve *partial* renewal, the stitching of cracks and the patching and matching of portions of walling and finishings, etc.

Each item should be SCHEDULED giving location and extent and be *deemed to include* pulling down and cutting away, hacking off, preparation, provision of new work, dubbing out and matching and marrying of new to existing, facings, facework, mouldings and ornament.

Note: Strengthening should be dealt with generally as described Externally under "Strengthening – Repairs"

.05

New Walls: Generally, these should be SCHEDULED as for "Renewal" and *deemed to include* for work as there defined, excluding taking down but including all cutting back to join up with existing construction and for preparation.

Where the work is *not* built of materials to match existing, their type and quality should be stated.

Note: For Depth of Foundations – see "Work Parameters – Principles"
For Finishes, Etc – see headings below
For Borrowed Lights, Doors, Etc – see headings below.

3.54 Floors

IN THE EVENT OF A WHOLE FLOOR BEING RENEWED, OR OF A NEW ONE BEING CONSTRUCTED, EXTENSIVE INCIDENTAL WORK IS INVOLVED: THE ORDER OF WORK SECTIONS AND PARAMETERS OF WORK FROM STRIPPING OUT TO REINSTATEMENT SHOULD GENERALLY FOLLOW THIS "MODEL".

IN RESPECT OF THE FLOOR ITSELF; WORK SHOULD BE UNDER SUB-HEADINGS AS FOLLOWS:

.01

Removal: *see "Pulling Down, Etc" p. 166.*

.02

Renewal: The SCHEDULE should state location and the extent of the work and include the taking out and renewal of joists and boarding/slab and screed, etc (depending on form of construction – which should be stated), also the requirements as to hangers, intermediate and end bearings, beams and other supports, encasing of beams, etc and the formation and finishing to openings.

The requirements of fire resisting grade and construction should be additionally defined.

Note: Regard must be had for Regulation scantlings – see "Work Parameters – General"

For Ceiling, etc Finishes – see headings below
For Floor Coverings – see headings below
For Skirtings – see headings below

.03

Isolated Openings in Existing Floors: Refers to the formation enlarging, reducing or the blocking up of isolated openings in floors. The SCHEDULE should give location(s) and size and be *deemed to include* cutting, adapting, forming or filling the opening, additional trimming members for the stability of the floor construction, for finishes at edges and the making out of floor and ceiling finishes where openings are filled, or partly filled in.

.04

Repairs: These will generally involve partial renewal and patching and matching of portions of flooring.

If isolated and located in a compartment or room, the work should be SCHEDULED there.

In respect of woodworm and/or dry rot, the extent should be fully determined and defined giving location(s) and extent of treatment.

If such treatment is to be carried out by a specialist a specification and firm quotation should be obtained and any separation of the respective work parameters established and defined.

Whether by builder or specialist, it should be ensured that the incidental cutting away, setting-aside and reinstatement of items to gain access for the work is additionally covered by them.

Note: Strengthening should be dealt with generally as described Externally under "Strengthening – Repairs".

.05

New Flooring: This could occur in an existing structure – to sub-divide a storey, or in a mezzanine, or in adding a storey but, as stated above, the work sections and parameters of the Model would apply to all the incidental work. In respect of the floor, the work should be SCHEDULED as for "Renewal" and *deemed to include* as there defined excluding taking out but including cutting back to join up with any existing construction and for preparation.

Where the work is *not* built of materials to match existing, their type and quality should be defined.

Note: For Ceiling, etc Finishes – see headings below
For Floor Coverings – see headings below
For Skirtings – see headings below

.06

Insulation: The requirements of sound insulation, grade and construction should be additionally defined.

3.55 Chimney Breasts

.01

Entire Removal: The SCHEDULE should state the location, number of flues, the number of storeys, size and height of the chimney breasts and of the chimney stack and be *deemed to include* for taking down the stack and chimney breast(s), cropping back and facing up the wall in brickwork block-bonded in, the removal of hearths and the making out of flooring, floor and ceiling finishes and roof construction and coverings, etc, and generally as for "Pulling Down, Etc" p. 166.

Note: *The removal of part only of a chimney breast should be dealt with in the compartment or room where this work is located – defining the form of support, or corbelling, in the schedule.*

.02

Renewal: The SCHEDULE should state location, the number of flues, the number of storeys and size and height of the chimney breasts and of the chimney stack and be *deemed to include* for taking down and rebuilding the chimney breasts and stack including cutting and block-bonding in, the forming of hearths, trimming of floor and roof construction, making out of floor and ceiling finishes, rough rendering through roof spaces, the formation of chimney gutters and flashings around the stack and of fireplace openings and hearths and the lining of flues.

Note: *Part renewal should be dealt with in the compartment or room where this work is located – defining the form of support, or corbelling, in the schedule.*

.03

New Chimney Breasts: The SCHEDULE should state location, etc as for "Renewal" and be *deemed to include* as there defined excluding taking down but including all cutting back to join up with existing and all preparation.

Where work is *not* built of materials to match existing, their type and quality should be stated.

.04

Repairs: These will be varied and will normally require to be defined and SCHEDULED in the compartment or room where the works are located.

3.56 Damp Proof Courses in Existing Walls

The SCHEDULE should define the method to be employed and extent of the installation stating the walls involved and be *deemed to include* all work to gain access and reinstatement of works disturbed.

Where a specialist firm is to be employed, a specification and firm quotation for the installation should be obtained and any separation of the respective work parameters established and defined – see "Work Parameters – General".

Whether by builder or specialist, it should be ensured that the incidental cutting away, setting aside and reinstatement of items to gain access for the work is additionally covered by them.

Note: Repairs arising from the lack of damp course should be included under "Repairs" in their respective rooms or compartments.

3.57 Compartments

AS DESCRIBED IN "WORK PARAMETERS – PRINCIPLES", COMPARTMENTS SHOULD FORM CLEAR CUT DIVISIONS OF A BUILDING, USUALLY SELF-CONTAINED; FOR EXAMPLE A HOUSE, A FLAT OR MAISONETTE, AN OPEN FLOOR, A COMMON STAIRCASE IN A FLATTED BLOCK, BEFORE FURTHER SUB-DIVISION INTO SPACES SUCH AS ROOMS.

THUS COMPARTMENTS SHOULD BE USED AS MAIN LOCATIONAL HEADINGS AND IF THEY OCCUPY SEVERAL STOREYS, THEY SHOULD BE SCHEDULED STOREY BY STOREY FROM TOP DOWNWARD.

3.58 Partitions

THESE USUALLY SERVE TO MAKE THE SUB-DIVISION TO ROOMS, OR THE LIKE AND ARE MORE CONVENIENTLY GROUPED UNDER A HEADING, EACH GIVEN ITS LOCATION.

ANY PARTITIONS TO BE REMOVED OR ALTERED SHOULD BE INCLUDED UNDER THIS HEAD.

.01

Partitions – Removal: The SCHEDULE should give location and extent and *be deemed to include* taking down, provision of supporting beam in lieu – if load bearing, cutting back at wall faces, etc and making out finishes to ceilings, walls and floors and generally as for "Pulling Down, Etc" above.

.02

Partitions – Renewal: The SCHEDULE should give location and extent and be *deemed to include* taking down the prescribed parts and provision of new including soleplates, studwork and the sub base(s) for finishes, blockwork, brickwork, etc (depending on form of construction – which should be stated), also work to openings, lintels, abutments and intersections and the setting-in of frames of borrowed lights, doors, etc. The requirements of fire-resisting grade and construction should be additionally defined.

Note: For Finishes, Etc – see headings below
For Borrowed Lights, Doors, Etc – see headings below

.03

Partitions – Isolated Openings: Refers to the formation, enlarging, reducing or the blocking-up of openings in existing partitions.

The SCHEDULE should give location and size and be *deemed to include* cutting, adapting, forming or filling the opening, lintels, finishes and features to head, reveals, cills and thresholds and setting in position and pointing of frames to borrowed lights, doors, etc and the making-out of finishes where openings are filled or partly filled in.

.04

Partitions – Repairs: These may include *partial* renewal and patching and matching of portions of partitions and finishings, etc.

Each item should be SCHEDULED and described giving location and extent and be *deemed to include* cutting away, preparation, provision of new work, dubbing out, matching and marrying of new to existing, facings, facework mouldings and ornament.

.05

Partitions – New: Generally, these should be SCHEDULED as for "Renewal" and *deemed to include* for work as there defined excluding taking down but including all cutting back to join up with existing and all preparation.

Where the work is *not* built of materials to match existing, their type and quality should be stated.

Note: For Finishings – see headings below
For Borrowed Lights, Doors, Etc – see headings below

.06

Partitions – Insulation: The requirements of sound and fire insulation, grade and construction should be additionally defined.

If insulation is provided by way of applied finishes, include under "Wall Plaster – Insulation" – later.

3.59 Rooms

WORK SHOULD BE SCHEDULED ROOM BY ROOM AND FOLLOW A SUITABLE TRADE SEQUENCE AS DESCRIBED UNDER "WORKS PARAMETERS – PRINCIPLES" AND AS DETAILED IN THE FOLLOWING PAGES:

ROOMS SHOULD BE SCHEDULED UNDER COMPARTMENT HEADINGS, STOREY BY STOREY, TOP DOWNWARD.

IN ORDER TO STANDARDISE THE NAMING OF ROOMS, IT IS SUGGESTED THAT THE TERM "LIVING ROOM" SHOULD APPLY TO A LOUNGE, PARLOUR, RECEPTION ROOM, DINING ROOM, ETC. AND THAT BEDROOMS AND LIVING ROOMS BE ABBREVIATED "BR" AND "LR" RESPECTIVELY WITH A NUMBER AFFIXED LOCATING THEM BY SIZE – THE LARGEST AS "1": HENCE, A THREE BEDROOM HOUSE WITH A LOUNGE AND DINING ROOM WILL BE ANNOTATED: BR1, BR2, BR3, BATHROOM, WC, KITCHEN (including Stores, Pantry, Etc), LR2, LR1, STAIRCASE (including Landings and Hall), 2nd WC (if any).

THERE WILL INVARIABLY ARISE A MISCELLANY OF MINOR ITEMS IN SCHEMES OF IMPROVEMENT, CONVERSION OR REPAIR; THESE SHOULD BE INCLUDED AS DETAILED UNDER 3.60 "MISCELLANEOUS SUNDRIES AND REPAIRS".

WHERE WORK IS COMMON TO SEVERAL ROOMS, FOR EXAMPLE CENTRAL HEATING, ITEMS SUCH AS RADIATORS AND OTHER RELEVANT PARTS OF THE INSTALLATION SHOULD BE NOTED WHERE THEY ARE LOCATED WITH CROSS REFERENCES TO ANY DETAILED ANNEXED SPECIFICATION.

.01

Fireplaces – General: These will require to be dealt with according to their situation, for example, bedrooms, living rooms and kitchens.

Where the fire combines a back-boiler (a) to heat water or (b) for central heating, the heating unit should be included later in the Hot Water and/or Central Heating Installation(s).

.02

Fireplaces – Blocking-in: The SCHEDULE should state type of fireplace and should be *deemed to include* taking out the surround and fire interior, sweeping flues, the brick block or stud, in-fill to the opening – which should be defined; also the provision of vent(s) to flue(s) and finishes to walls and floor.

Note: For Work to Stack – see "Externally – Sundries"

.03

Fireplaces – Renewal: The SCHEDULE should state the type and quality of the new fire interior, surround and hearth and be *deemed to include* for taking out existing, preparing for and installing new including fire lintel, backing and provision of all new works to complete their installation.

Note: Where the interior is a combined fire and back boiler and is to be included under Hot Water and/or Central Heating Installation, the installation of surround and hearth are to be included here, with a cross-reference inserted about the fire and back boiler elsewhere.

For Chimney Breasts – see "Internally'

.04

Fireplaces – New: Generally these should be SCHEDULED as for "Renewal" and be *deemed to include* for work as there defined excluding taking out existing but including all cutting away and preparation.

Note: For Chimney Breasts – see "Internally"

.05

Wall Plaster, Etc – General: This work may arise as a consequence of other work such as structural adaptations, extensions, etc.

Repairs should be kept separate under the heading "Repairs" and minor items may be grouped under a heading "Sundries".

.06

Wall Plaster, Etc – Renewal: The SCHEDULE should state if to match existing and give extent, otherwise it should define type, class of finish and quality and be *deemed to include* hacking off or stripping existing, cutting back for key, preparation, dubbing out, replastering, relining or retiling, matching and marrying new to existing facework, mouldings and ornament, arrises and edges.

The requirements of fire resisting grade and construction should be additionally defined.

.07

Wall Plaster, Etc – New: The work should be SCHEDULED as for "Renewal" and *deemed to include* for works as there defined excluding hacking off or stripping existing but including cutting back for key and preparation.

.08

Ceiling Plaster – Renewal: Generally as for "Wall Plaster" above.

.09

Ceiling Plaster – New: Generally as for "Wall Plaster" above.

.10

Screeds – Renewal: Generally as for "Wall Plaster" above.

.11

Screeds – New: Generally as for "Wall Plaster" above.

.12

Insulation: The requirements of sound and fire insulation, grade and construction by applied finishes should be additionally defined.

.13

Flooring – General: Any major item of work will already have been dealt with if structural under "Floors" above.

Incidental work arising from structural work to floors, such as skirtings, should be defined and scheduled under the heading "Skirtings, Etc" below.

Isolated repair items of flooring in rooms now arising should be defined and scheduled under "Repairs" below.

Miscellaneous items should be defined and scheduled under "Sundries" below.

In respect of woodworm and dry-rot, where these are isolated or distinctly located in a compartment or room, the work should be defined and scheduled there.

The requirements of fire resisting grade and construction should be additionally defined.

Note: Floor boarding and screeds will have been composited with the construction if dealt with under 3.54 "Floors" above.

.14

Flooring – Coverings – Renewal: These may include applied or laid-over finishes in addition to the first or basic floor finish, for example, lino over boarding, tiles on screed, etc. They may also include an underlining such as hardboard; or simply planing and sanding.

The SCHEDULE should define extent, type and quality and state if to match existing and be *deemed to include* taking up existing, preparing surfaces and laying new, any pattern and margin, all work to edges, etc and cutting and fitting of every description.

.15

Flooring – Coverings – New: Generally, these should be SCHEDULED as for "Renewal" and *deemed to include* for work as there defined excluding taking up existing.

.16

Flooring – Insulation: The requirements of sound and fire insulation grade and construction by applied finishes should be additionally defined.

.17

Windows (and Frames) – Renewal: There may be alternative compositions which may include the sashes or frame only, or sashes and frame, or sashes, frame and fanlight, etc.

The SCHEDULE should define extent of renewal of these parts and state if to match existing; otherwise give, size, type and quality and be *deemed to include* taking out and refixing or rehanging. The provision of new sashes, fanlights, frames, architraves, etc, ironmongery and glass as required, all piecing – in, pointing and making good.

.18

Windows (and Frames) – New: Generally, these should be SCHEDULED as for "Renewal" and be *deemed to include* for all new work, as there defined excluding taking out but including all cutting away and preparation.

.19

Doors (and Frames) – Renewal: There may be alternative compositions which may include the door or frame; or the door and frame; or door, frame, fanlight etc.

The SCHEDULE should define the extent of renewal of the various parts and state if to match existing; otherwise give size, type and quality and be *deemed to include* taking out and refixing or rehanging, the provision of new doors, fanlights, frames, architraves, etc ironmongery and glass as required, all piecing-in, pointing and making good.

The requirements of fire resisting grade and construction should be additionally defined.

.20

Doors (and Frames) – New: Generally, these should be SCHEDULED as for "Renewal" and be *deemed to include* for all new work, as there defined excluding taking out but including all cutting away and preparation.

.21

Borrowed Lights (and Frames) – Renewal: Generally as for "Doors – Renewal" above.

.22

Borrowed Lights (and Frames) – New: Generally, as for "Doors – New" above.

.23

Cupboards, Kitchen Fittings, Etc – Renewal: These should include sundry fittings and shelves.

The SCHEDULE should define their composition and state if they are to match existing otherwise give size, type, quality and location and be *deemed to include* for taking out existing and provision of new and ironmongery, glass and trim as required, all piecing-in and making good.

Note: Work to different types of cupboard should be identifiable such as to Pantry cupboards, Store cupboards, Linen cupboards, and to Kitchen fittings and similar types classed as manufactured joinery.

.24

Cupboards, Kitchen Fittings, Etc – New: Generally, these should be SCHEDULED as for "Renewal" and be *deemed to include* for work as there defined excluding taking out but including all cutting away and preparation.

Note: as for "Renewal" above.

.25

Skirtings, Dadoes, Picture Rails, Trim and the like – Renewals: These are items which arise mainly as a consequence of other work such as structural alterations, filling-in openings, etc but *not* as repair, which should be kept separate under "Repairs"; or as a group of minor incidental items under "Sundries".

The SCHEDULE should state extent and if to match existing otherwise size, type and location should be given and be *deemed to include* taking out existing work and preparing for and fitting new including matching, cutting and fitting of every description and making good.

.26

Skirtings,Dadoes, Picture Rails, Trim and the like – New: Generally, these should be SCHEDULED as for "Renewal" and be *deemed to include* preparing for and fitting new including matching etc all as there described.

.27

Stairs – General: These comprise the stairs, landings and balustrades within the staircase compartment.

They should be located by the storey(s) in which the stairs are sited; for example, Ground Floor when rising from Ground to First Floor, and so on.

.28

Stairs – Renewal: The SCHEDULE should define the extent to which an existing staircase is to be taken out and renewed and if to match existing, otherwise give size, type, quality and location stating storey height(s) and going(s), the construction of treads, risers, landings, strings, bearers, newels, handrails, balusters, aprons, soffites, spandril in-fillings and shall be *deemed to include* taking out to the extent defined and preparing for and fitting new including matching, cutting and fitting of every description and making good.

The requirements of fire-resisting grade and construction should be additionally defined.

.29

Stairs – New: Generally, these should be SCHEDULED as for "Renewal" and be *deemed to include* for work as there defined excluding taking out but including cutting away and preparation.

.30

Sanitary Fittings – General: Baths, Basins, Sinks, WC Suites and other sanitary fittings should each be regarded as composite units complete with taps, traps, over-flows, soil and waste pipes (to point of discharge at drain, gulley or stack pipe) and for connecting pipes (or "tails") to hot and cold supplies.

Where sinks have drainers and/or cupboard units in combination and baths are enclosed, or where any fitting has other subsidiary parts, these should be defined.

.31

Sanitary Fittings – Renewal: Unless all parts are similar to existing, they should be fully described.

The SCHEDULE should define the composition stating size, type, quality and location and be *deemed to include* for cutting-off of supplies, dismantling and taking out existing, assembly of new, the pipework, fittings, joints and fixings, all cutting away and reinstating as required in carrying out the installation, testing and leaving in sound working order.

.32

Sanitary Fittings – New: Generally, these should be SCHEDULED as for "Renewal" and *deemed to include* for work as there defined excluding dismantling and taking out existing but including cutting back to join up with any existing services and for preparation.

3.60 Miscellaneous "Sundries" and "Repairs"

THESE SHOULD BE SCHEDULED AT THEIR RESPECTIVE LOCATIONS AND GROUPED IN THE COMPARTMENTS OR ROOMS WHERE THEY OCCUR EITHER AS SUNDRIES, OR REPAIRS, AS FOLLOWS:

.01

Sundries: As stated in "Work Parameters – Principles" there will inevitably be miscellaneous items of work *other than repairs* that will need to be carried out to complete works of improvement or conversion and additionally to meet the full requirements of the client.

Examples of such work, in Trade sequence, may be:

Brickwork, etc in connection with any such items.
Splash-backs.
Overlay to a basic floor finish.
Alterations of windows to change the style.
Alterations of doors to change the style.
Provision of fitted bedroom cupboards
Changes of style in picture rail, etc or their removal.
Provision of bathroom, toilet, etc fittings, eg mirrors.

Extension of water service for garden or garage use.
Provision of H & C supplies in excess of "Parker Morris".
Full central heating (ie excess of "Parker Morris").
Provision of gas cookers.
Provision of electric cookers etc and/or points in excess of "Parker Morris".

Garages and Drives, Garden Works, Etc.

.02

Repairs: Attention is drawn to "Works Parameters – Principles" – "Repairs" and the more precise definition adopted for the Method than the everyday use of the term. In short, the Method applies the term "Repairs" generally to minor items relating to major parts or elements; for example, to the "stitching" of a crack in brickwork, the renewal of a sash to a window or of items of ironmongery and so on.

Note: Whilst renewal is used in respect of certain repairs as in common usage (see examples below), the term "renewal" is generally applied in the Model to more major parts or elements.

Examples of such work, in Trade sequence, may be:

Renewal of fireback, basket and fret.
Cutting out and making good loose plaster.
Cutting out and making good defective flooring.
Renewing sash and/or sash cords.
Easing door and/or renewing lock and furniture
Refixing shelves.
Cutting out and piecing – in existing sections of skirting.
Rewashering, or renewing taps to both.

Renewing stop valve to incoming water main.
Renewing storage tank, ball valve, overflow and insulation.
Renewing flue lining.
Refixing loose gas service piping.
Refixing ceiling rose and renewing flex to light pendant.

Renew broken balusters to stairs.

Note: Maintenance is considered to be repair of an ongoing nature such as occasioned by ordinary wear and tear and redecoration including the incidental making good of surfaces in preparation. Maintenance when coinciding with works primarily of repair may need to be kept separate under a heading "Maintenance".

3.61 Services

SERVICE INSTALLATIONS WILL NORMALLY HAVE COMMON MAINS THAT SERVE MORE THAN ONE COMPARTMENT OR ROOM, AND MAY ALSO BE THE SUBJECT OF SEPARATE CONTRACTS; EACH SERVICE IS THEREFORE MORE APPROPRIATE GIVEN UNDER A SEPARATE HEADING, AS SHOWN HEREUNDER:

3.62 Cold Water Supply

.01

Cold-Water Supply – Renewal: This may extend from the water authority's main to the stop-cock first reached inside the building and be sub-divided to (a) works outside, and (b) works inside the property boundary.

The SCHEDULE should define if the installation is to be as existing, otherwise it should state the nature, quality and extent and be *deemed to include* cutting-off of supplies, dismantling, the taking out of existing and assembly of new, all pipework, fittings, valves, joints, insulation and fixings, trenches for pipes if underground, all cutting away and reinstatement required in carrying out the installation, testing and leaving in sound working order.

In respect of work outside the boundary, that required to be carried out by the local authority and water authority should be separately defined.

.02

Cold Water Supply – New: Generally, this should be SCHEDULED as for "Renewal" and *deemed to include* for work as there described excluding dismantling and taking out existing but including cutting back to join up with any existing supply and for preparation.

3.63 Hot and Cold Water Installation

.01

Hot and Cold Water Installation – General: This should be sub-divided to:

(a) Rising Main (from stop cock first reached inside building to cold water storage tank);

(b) Cold Water Storage tank;

(c) Cold supplies from rising main and down service to serve fittings;

(d) Hot water cylinder;

(e) Hot supplies to serve fittings;

(f) Heat source – which may be an immersion Heater and/or Flow and Return pipes (either from a separate Hot Water Boiler/Water Heater or Central Heating Boiler and

(g) separate Hot Water Boiler/Water Heater.

Note: For connections (or "tails") from hot and cold supplies to fittings, see "Sanitary Fittings".

.02

Hot and Cold Water Installation – Renewal: The SCHEDULE should define the extent of works covering the various headings above and state if to match existing, otherwise to state type, class, quality and extent and be *deemed to include* cutting-off of supplies, dismantling and taking out existing and installation of new including tank, cylinder, heater/boiler/heat exchanger, thermostats, pumps, all pipework, duct-work, fittings, bosses, valves, overflows, joints and fixings, insulation, flues and flue linings, related electrical circuit and fittings, all cutting away and reinstatement required in carrying out the installation, testing and leaving in sound working order.

Note: For Chimney Breasts – see Internally.

.03

Hot and Cold Water Installation – New: Generally, this should be SCHEDULED as for "Renewal" and be *deemed to include* for work as there described excluding dismantling and taking out existing but including cutting back to join up with any existing installation and all preparation.

Note: For Chimney Breasts – see Internally.

3.64 Central Heating Installation

.01

Central Heating Installation – General: The work content will depend on the system of heating which may be one complete system or a combination of various alternatives and may include the heating of the hot water supply.

The heating requirements of the installation should be defined.

.02

Central Heating Installation – Renewal: The SCHEDULE should state if the installation is to be as existing otherwise the extent of requirements should be defined giving the system(s) to be employed such as Block Storage, Under floor, Warm Air, Back or Independent Boiler combined with Hot Water Radiators, Etc and, state if based on Electricity, Gas, Oil or Solid Fuel and give location, size, type, capacities and relative particulars of all parts, including associated ventilators, water supply and electrical circuits and fittings and be *deemed to include* cutting-off of supplies, dismantling and taking out existing and installation of new including tanks, cylinders, heaters/boilers/heat exchangers, thermostats, emitters, ventilators, all pipework ductwork, fittings, bosses valves, overflows, joints and fixings, insulation, flues and flue linings, related water supply and electrical circuits and fittings, all cutting away and reinstatement required in carrying out the installation, testing and leaving in sound working order.

The work of connecting to service mains and the requirements of fuel storage should all be separately defined.

Where a specialist firm is to be employed, a specification and firm quotation for the installation should be obtained and any separation of the respective work parameters established and scheduled – see "Work Parameters – General".

Where a specification is annexed to the schedule of works, the siting of heating points, etc. should be SCHEDULED or shown on drawings and cross referenced to the specification – see "Works Parameters – General".

Note: For Chimney Breasts – see Internally.

.03

Central Heating Installation – New: Generally, this should be as for "Renewal" and be *deemed to include* for work as there described excluding dismantling and taking out existing but including cutting back to join up with any existing installation and all preparation.

Note: For Chimney Breasts – see Internally.

3.65 Gas Installation

.01

Gas Installation – General: The work content will depend on requirements and whether it is an entire or partial installation, also whether it is split between several consumers.

.02

Gas Installation – Renewal: The SCHEDULE should state if the installation is to be as existing, otherwise the points of supply should be defined including metering and mains connection and be *deemed to include* cutting-off of supplies, dismantling and taking out existing and installation of new including tubes and tubulers, tees, elbows, flanges, caps, cocks, sleeves, clips and fixings, etc, all cutting away and reinstatement required in carrying out the installation, testing and leaving in sound working order.

The work of connecting to Gas Board's main should be separately defined.

Where the Gas Board or specialist is to be employed, a specification and firm quotation for the installation should be obtained and any separation of the respective work parameters established and defined – see "Work Parameters – General".

Where a specification is annexed to the Schedule of Works, the siting of gas points should be SCHEDULED or shown on drawings and cross referenced to the specification – see "Works Parameters – General".

Note: For Chimney Breasts – see Internally.

.03

Gas Installation – New: Generally, this should be SCHEDULED as for "Renewal" and be *deemed to include* for work as there described excluding dismantling and taking out existing but including cutting back to join up with any existing installation and all preparation.

Note: For Chimney Breasts – see Internally

3.66 Electrical Installation

.01

Electrical Installation – General: The work contract will depend on requirements, whether it is an entire or partial installation and if split between separate consumers.

The Lighting and Power requirements should be separately defined for each consumer.

.02

Electrical Installation – Renewal: The SCHEDULE should state if the installation is to be as existing, otherwise the extent of requirements should be defined giving the location, size, types, capacities, and relative particulars of all parts including metering and mains connections and be *deemed to include* cutting-off of supplies, disconnecting and stripping out existing and installation of new including wiring, conduit, covers, boxes, outlets, switches, pendants, lampholders, insulation, fixings, control and consumer units, all cutting away and reinstatement required in carrying out the installation, testing and leaving in sound working order.

The work of connecting to the Electricity Boards meters and mains should be separately scheduled.

Where the Electricity Board or specialist is to be employed, a specification and firm quotation for the installation should be obtained and any separation of the respective work parameters established and defined – see "Work Parameters – General".

Where a specification is annexed to the schedule of works, the siting of electrical points etc. should be SCHEDULED or shown on drawings and cross referenced to the specification – see "Works Parameters – General".

.03

Electrical Installation – New: Generally, this should be SCHEDULED as for "Renewal" and be *deemed to include* for work as there described excluding disconnecting and stripping out existing but including cutting back to join up with any existing installation and all preparation.

3.67 Decorations

.01

Decorations: These should be SCHEDULED to define type, quality, extent and location – by room or compartment – in order of ceilings, walls, woodwork, metalwork, etc and be *deemed to include* the bringing forward of surfaces and preparation for painting.

Standard definitions – drainage, etc.

3.68 Drainage, Outbuildings, Works Around the Buildings and to Boundaries

.01

Drainage – General: The work content will depend on whether it concerns an entire system or part of a system and should include soil, waste and vent stacks.

The system should be sub-divided to:

(a) Soil, waste and vent stacks;

(b) Branch drains and gullies

(c) Soakaways;

(d) Manholes;

(e) Main drain;

(f) Interceptor;

(g) Sewer connection, cesspool or septic tank.

.02

Drainage – Renewal: The SCHEDULE should define the extent of work covering the various headings above and state if to be as existing, otherwise to state, size, type, class, quality and extent and be *deemed to include* disconnection and removal of existing, excavation, support and back-fill to trenches, installation of new drainage including pipes, gulleys, raising pieces, double collars, plugs, traps, bends, junctions, channels, beds, haunching and encasing as required, soakaways, manholes, covers, vents, all cutting away and reinstatement in carrying-out the installation, testing and leaving in sound working order.

The sewer connection should be separately defined.

Cesspools and septic tanks should be separately defined.

.03

Drainage – New: Generally, this should be SCHEDULED as for "Renewal" and be *deemed to include* for work as there described excluding disconnection and removal of existing but including cutting back to join up with any existing and all preparation.

.04

Outbuildings, Etc: This may include Garages; Drives; Pavings; Garden Works; Boundary Walls; Fences, Etc.

The work should be SCHEDULED and defined generally in accordance with the Method.

Glossary of terms

Note: A Section on "The Usage of Sundry Building Terms" is to be found at the end of Section 2.

Many words relating to documentation practice and procedure for improvement and conversion works are casually used, yet they can usefully be given distinctive meanings. Words most commonly used, additional to Rehabilitation commonly used for housing, are:

> Reinstate, Renovate, Restore, Recondition, Renew, Repair, Refurbish, Replace and Improve.

The particular meanings given below have been found necessary in analysing documentation for improvement and conversion works from very many sources, private and public.

But before making those distinctions, it might be of interest to note what Sir Ernest Gowers, in 1954, made of the word "Rehabilitate". In his book "The Complete Plain Words" he firstly quoted Ivor Brown saying:

> *"The present darling of the Departments . . . is rehabilitation, a word originally applied to the restoration of a degraded man's rank and privileges. By the middle of the nineteenth century it was occasionally used to mean restoration of other kinds. Suddenly it has become the administrator's pet. A year or two ago nothing was mended, renewed or restored. Everything had to be reconditioned. Now reconditioning has been supplanted by rehabilitation, which has the merit of being one syllable longer; the blessed word "goes" officially with everything from houses to invalids. I can see no reason why the Ministry of Health should not still seek to heal people instead of rehabilitating them. But heal – poor old Bible monosyllable! Will the next translation of the Bible be allowed to heal the sick? No, it will have to rehabilitate those who are suffering from psycho-physical maladjustment."*

To this statement Sir Ernest added:

> *"But it is only fair to remark that rehabilitation, thus used, means something more than healing. It means a course of treatment or instruction for the purpose of restoring people already healed of a disease or wound to a life of active usefulness. Because this extension of the healing art was a new*

conception, it was given a new name, reasonably enough, however ill-chosen the name may be thought to be. What is to be deplored is that "the blessed word goes officially with everything from houses to invalids."

The manner in which particular meanings can be shown diagrammatically is as follows:

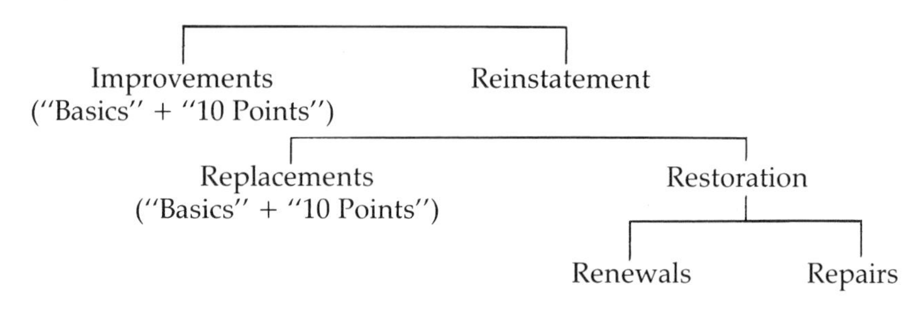

The prime objectives are to improve and repair. Broadly, the former item implies adding something new and the latter to reinstate that which exists to its proper condition.

But improvements prescribed by the Housing Acts are laid down in the 1980 Act (see extracts appended to this Glossary): when provided for the first time these are regarded as "improvements" but if reinstated, the term "replacement" is used.

Repair, as stated earlier is a commonly used term and implies renewal in different parts of a building in varying degree. The term "restoration" is therefore used to embrace all repair including renewal so each can have a specific meaning: "Renewal" is therefore applied to the restoration of major parts or elements and "Repair" to *minor parts* of major parts or elements.

It has also been found necessary to apply particular meaning to certain other terms, such as Alteration, Adaptation and Extension: "Alteration" is used to embrace both Adaptations and Extensions; "Adaptation" applying to changes (primarily structural) within the existing envelope of the building and "Extension" to structural additions built-on.

Various terms with their particular applications follow:

ADAPTATIONS: Changes, usually structural, within the existing envelope of the building.

ALTERATIONS: Covers changes, usually structural, by ADAPTATION (ie within the envelope) *and* EXTENSION (ie additions built-on).

AMENITIES: Contribute to the standard of accommodation by including those laid down by the Housing Act 1974 ie the five "basic" amenities; the "10 points" and "repair". See HOUSING ACT REQUIREMENTS following this Glossary.

BASICS: See the STANDARD AMENITIES under Housing Act requirements at the end of this Glossary.

CONVERSION: Considered to refer to alteration resulting in a change in the containment of space in a building *and* its use; such as the change from a house into self-contained flats, or vice versa, usually with improvements (ie addition of amenities) and repair (ie restoration to sound condition).
Includes IMPROVEMENT, REPLACEMENT, RENEWAL, REPAIR and ALTERATION.

ELEMENT: Regarded as a readily identifiable composite part of a building such as a wall, floor, roof, door, window, etc. The extent of their respective compositions (ie their "Parameters") being defined in Sec 3.

EXTENSION: Taken to mean a structural addition to the envelope built onto an existing building.

IMPROVEMENTS: Generally includes the addition of amenities plus repair and may include alterations but particularly applied, the term is limited to cover the provision *for the first time* of the five "BASICS" *and/or* the "10 POINTS" *and* may include ALTERATIONS (ie structural changes).

MAINTENANCE: Periodic reinstatement to sound and stable condition to remedy the effects of ordinary wear-and-tear including REPLACEMENT of obsolete parts and RESTORATION by renewal, repair, cleaning down, repainting and redecoration.

MAKING GOOD: In common usage this is loosely applied in a similar way to repair but more specifically is applied to work consequent on the carrying out of other work.

MODERNISATION: Largely includes changes to update standards in quality, style and accommodation.

REINSTATEMENT: Generally refers to the putting back of an item or element in sound and stable condition, much the same in form and position as previously existing. Particularly applied, the term is limited to cover REPLACEMENT (ie reinstatement of any of the five "Basics" or the "10 points") *and/or* RESTORATION (ie RENEWAL or REPAIR) including dismantling, pulling down, taking out, etc.

RENEWAL: Generally applies in a similar way to reinstatement but particularly applied, the term is limited to cover the RESTORATION of a major part or element of a building including dismantling, pulling down, taking out, etc.

REPAIR: Generally applies in a similar way to reinstatement but particularly applied, the term is limited to cover the RESTORATION of *a minor part* of a major part or element of a building including dismantling, etc.

Note: Within the context of repair the term renewal may of necessity be loosely used, eg renewal of a sash fastener to a window.

REPLACEMENT: Generally applies in a similar way to reinstatement but particularly applied, the term is limited to cover the reinstatement of any of the five "BASICS" and the "10 POINTS" including dismantling, etc.

RESTORATION: Generally, like reinstatement, this term is loosely used but particularly applied, it is limited to RENEWAL (referring to major parts or elements) *and/or* REPAIR (referring to *minor parts* of major parts or elements).

SUNDRIES: In common usage applies to a miscellany of items but given a more limited application the term may cover minor parts other than repair; also, items of making good (ie works consequent on the carrying out of another piece of work) and works of a more substantial though relatively less frequent kind eg lifts.

THE "10 POINTS" See items (1) to (10) listed under the Housing Act '80 requirements at the end of this Glossary.

The above list is in no way exhaustive of words used in documentation; other words commonly used could include revitalise, regenerate and remodel. It has not however been found necessary to use these in any special way.

Several terms relating to tendering, contracting, pricing and costs are also relevant to documentation but are not included because they are already more precisely defined in guides to contracting procedures.

"Repair" is commented upon in Section 1 of the Manual to the building regulations 1985 as follows:

> **"The Regulations say nothing about the point at which repair which is not subject to control becomes work which has to be controlled. That is a judgement which has to be made and depends on circumstances. Repair is basically replacement, or making good, and not new work or alteration.**
>
> **But in the case, for example, of whole building which has been seriously damaged, there comes a point when so much has to be done to repair or replace it that the local authority could reasonably require it to be treated as if a new building was being erected. In such a case the Regulations apply. If you are in doubt the local authority may be able to advise you."**

Legislation

Requirements under the Housing Act 1980

The following extracts from Circular 21/80 (Welsh Office Circular 42/80) gives the Directions and Specifications relating to Part VII of the Housing Act 1974 as amended by the Housing Act 1980 and which, if provided in existing houses, may attract grants under the Act:

THE "10 POINTS" and STANDARD AMENITIES

Appendix A, para 5

5. In exercise of his powers under subsection (3)(c) of section 61 of the 1974 Act, the Secretary of State *hereby specifies* that the requirements with respect to construction and physical conditions and the provision of services and amenities with which a dwelling must conform on completion of works of improvement or conversion (in addition to the other standards laid down in section 61(3) but subject to the powers of local authorities in section 61(4) and (4A) to dispense with any requirement to the extent they think fit) are that the dwelling must:–

(1) be substantially free from damp;

(2) have adequate natural lighting and ventilation in each habitable room;

(3) have adequate and safe provision throughout for artifical lighting, and have sufficient electric socket outlets for the safe and proper functioning of domestic appliances;

(4) be provided with adequate draining facilities;

(5) be in a stable structural condition;

(6) have satisfactory internal arrangement;

(7) have satisfactory facilities for preparing and cooking food;

(8) be provided with adequate facilities for heating;

(9) have proper provision for the storage of fuel (where necessary) and for the storage of refuse; and

(10) have, in the roof space, thermal insulation sufficient to give, for the relevant structure, a U value of 0.4 $W/m^{2}{}^{\circ}C$, where–

"relevant structure" means the roof over the roof space and the ceiling below it, together with the insulating material provided in the roof space, and

"U value" means thermal transmittance coefficient, that is to say the rate of heat transfer in watts through one square metre of a structure when the combined radiant and air temperatures at each side of the structure differ by one degree Celsius and is expressed in watts per square metre of surface area per one degree Celsius of temperature difference ($W/m^{2}{}^{\circ}C$).

Appendix B, para 54

54. The *standards* specified by the Secretary of State in exercise of his powers under section 61(3)(c) (see Appendix A) incorporate those matters specified in section 4 of the Housing Act 1957 (which relates to "fitness") that are not covered by the provisions of section 61(3)(a) and (b) of the 1974 Act. The references to, for example, structural stability and internal arrangement should be interpreted in accordance with the standard adopted for the purposes of section 4 of the Housing Act 1957. It is not appropriate for local authorities to seek to impose higher standards (eg Parker Morris standards), although if an applicant carrying out a comprehensive scheme of improvement proposed a standard comparable to Parker Morris in any particular respect he could be assisted with grant. Authorities are reminded that separate grants are available under the Homes Insulation Scheme to enable dwellings with no loft insulation to be brought up to standard in that respect; improvement grants should therefore not be used to aid such work.

THE STANDARD AMENITIES are listed in Part 1 of Schedule 6 to the 1974 Act:

A fixed bath or shower
A wash hand basin
A sink
A hot and cold water supply to each of the above items
A water closet.

STRUCTURAL STABILITY: References to Sec 4 of the Housing Act 1957 state:

Evidence of instability is only significant if it indicates the probability of further movement which would constitute a threat to the occupants of the house.

INTERNAL (BAD) ARRANGEMENT:

Is any feature which prohibits the safe unhampered passage of the occupants in the dwelling, eg narrow, steep or winding staircases, absence of handrail, inadequate landings outside bedrooms, ill defined changes in floor levels, a bedroom entered only through another bedroom and also includes a WC opening directly from a living room or kitchen.

Requirements under the Local Government & Housing Act 1989

The following summary is based on the provisions of the above Act that received the Royal Assent on 16 November 1989.

The foregoing requirements under the Housing Act 1980 are replaced by the passing of the Local Government & Housing Act 1989, whereby the new "fitness" standard is based on the requirements of section 604 of the Housing Act 1985 (per schedule 9) of the '89 Act – see below).

Local councils (as local housing authorities) have a duty to approve grants for dwellings, or houses in multi-occupation (HMOs) – in the main – which do not reach this standard.

There will continue to be grants for repairs and adaptations for the benefit of disabled people; for the common parts of buildings containing flats that will additionally qualify for means of escape from fire and associated works; for sundry minor works including insulation, and for houses in groups. Discretionary grants will be approved for those that are not mandatory or for works that may be additional to those necessary to make a property "fit"; for these there is an alternative standard (per section 115 of the Act – see below).

In respect of area improvement, the two former housing action areas (HAAs) and general improvement areas (GIAs), are replaced by one of renewal areas (RAs).

Grants will be made on the application of owner occupiers, tenants and landlords who lack sufficient resources themselves and whose premises are considered to be below a basic *fitness standard*. As well as needing the installation of amenities like baths, basins, sinks, hot and cold water and WCs (as in cases supported by intermediate grants hitherto), all aspects of the standard must be considered. For tenants to qualify as applicants for grants they will normally have to be liable for repairs, although they can apply for a disabled facilities grant; landlords will be restricted to mandatory grants on statutory repair notices, otherwise both will be confined to discretionary grants.

Certificates of future occupation will still need to be supplied by applicants as appropriate.

Authorities will be expected to require at least two estimates and when making grants will need to distinguish clearly those works (whether mandatory or discretionary) that are eligible for grant.

Authorities will be able to initiate group repair schemes (superceding the previous "block" and "enveloping" arrangements) and will be limited to statutory housing renewal areas.

Under this new regime, eligibility for grant will no longer be decided by rateable value (RV) limits; instead, determination of grant will depend on an applicant's ability to finance the work, or make a financial contribution towards it unless on benefit. This will mean enquiring into the applicants resources on the one hand, or landlord's ability to recover the cost of works from rental income on the other.

MANDATORY "FITNESS" STANDARD per section 112 of the '89 Act "to render certain dwellings fit for human habitation". A dwelling is expected to be fit when:

(a) it is structurally stable;
(b) it is free from serious disrepair;
(c) it is free from dampness prejudicial to the health of the occupants (if any);
(d) it has adequate provision for lighting, heating and ventilation;
(e) it has an adequate piped supply of wholesome water;
(f) there are satisfactory facilities in the dwelling for the preparation and cooking of food, including a sink with a satisfactory supply of hot and cold water;
(g) it has a suitably located water closet for the exclusive use of the occupants (if any);
(h) it has, for the exclusive use of the occupants (if any), a suitably located fixed bath or shower and wash hand basin, each of which is provided with a satisfactory supply of hot and cold water; and
(i) it has an effective system for the draining of foul, waste and surface water

DISCRETIONARY STANDARD per section 115 of the '89 Act "of certain applications". Approval may be made when:

Under subsection (1):

(a) the relevant works go beyond or are other than those which will cause the dwelling to be fit for human habitation, but
(b) the authority are satisfied that the relevant works are necessary for one or more of the purposes set out in subsection (3) below.

(3) The purposes referred to in subsection (1) above are:

- (a) to put the dwelling or building in reasonable repair;
- (b) to provide the dwelling by the conversion of a house or other building;
- (c) to provide adequate thermal insulation;
- (d) to provide adequate facilities for space heating;
- (e) to provide satisfactory internal arrangements;
- (f) to ensure that the dwelling or building complies with such requirements with respect to construction or physical condition as may for the time being be specified by the Secretary of State for the purposes of this section; and
- (g) to ensure that there is compliance with such requirements with respect to the provision or condition of services and amenities to or within the dwelling or building as may for the time being be so specified.

Under subsection (2) in respect of common parts of buildings, works:

- (a) are necessary for one or more of the purposes set out in paragraphs (a) and (c) to (g) of subsection (3); or
- (b) will cause the building to meet the requirements mentioned in paragraphs (a) to (e) of section 604(2) of the Housing Act 1985, namely:
  - (a) the building or part is structurally stable;
  - (b) it is free from serious disrepair;
  - (c) it is free from dampness;
  - (d) it has adequate provision for ventilation; and
  - (e) it has an effective system for the draining of foul, waste and surface water.

Note 1: *All the requirements are subject to all the related provisions of the Act.*

Note 2: *The standards required by earlier legislation are retained in the text as the transition to the above standards is taking place when this review is being prepared.*

Section Four

Contract

4.00 Forms of Contract and Notes for Guidance in the Selection of a Form to use on "Small" or "Minor" works.

.01

Attached are tabulated the clause headings of recommended contract forms so that their respective provisions can be readily compared: the short titles of the forms are:

"STANDARD FORM" of Building Contract, 1980 Edition.

AGREEMENT FOR MINOR BUILDING WORKS, 1980 issue.

AGREEMENT FOR RENOVATION GRANT WORKS, December 1975 Revision.

AGREEMENT FOR RENOVATION GRANT WORKS, where no architect/supervising officer is appointed, December 1975 Revision.

INTERMEDIATE FORM of Building Contract 1984 Edition.

(Other forms are produced by certain professional organisations. Builders run: The Building Employers' Confederation "GUARANTEE SCHEME" and The Federation of Master Builders "WARRANTY SCHEME").

.02

All the above JCT contract forms are for use on works proceeding on drawings and/or specification.

.03

The STANDARD form is produced:
"without quantities", "with quantities" and "with approximate quantities".

.04

The STANDARD forms are additionally published in editions for private and local authority use.

.05

The STANDARD form (without quantities) provides for the use of schedules of rates.

.06

The notes and the tabulations are presented to assist in choosing a contract form appropriate to the job requirements and indicate the measure of control provided.

.07

The editions are those currently available as at date of publication but editions of forms should be checked that they are up to date. The dates of the issue of those now current are shown above and in the headings to the tabulations. Amendments published from time to time should also be used.

.08

The MINOR WORKS Form and the GRANT Forms are not for use in Scotland.

.09

The several JCT forms are issued by the Joint Contracts Tribunal, through their publisher, RIBA Publications Ltd, Finsbury Mission, Moreland St, London ECIV 8VB.

.10

Information about the Builders' Schemes is given later, also the addresses where certain professional organisations forms can be obtained.

.11

All forms are designed for use where an architect/supervising officer is employed EXCEPT THE GRANT form in the last column and "private" versions of the Standard Form where Supervising Officer is omitted.
Note: Supervising Officer applies to a person who is not a registered architect. "Private" distinguishes forms so described from those for "Local Authority" use.

.12

The NOTES for Guidance last issued by the Joint Contracts Tribunal concerning the GRANT Forms are reproduced (by their kind permission) following the tabulations.

.13

Each form carries articles of agreement for completion by the parties.

.14

Architect/Supervising Officer Delete where appearing on the forms, whichever of these is not applicable.

.15

The Intermediate Form of Contract 1984 is described as a 'Form of building contract for works of simple content' and Practice Note 20 (revised 1984) should be read in conjunction with it. The JCT describes the use of the Intermediate Form as follows:

> "This Intermediate Form is issued for contracts in the range between those for which the JCT Standard Form of Building Contract with quantities (1980 Edition) and the JCT Agreement for Minor Building Works (1980 Edition) are issued.
>
> The Form would be suitable where the proposed Building Works are:
>
> 1. of a simple content involving the normally recognised basic trades and skills of the industry; and
>
> 2. without any building service installations of a complex nature, or other specialist work of a similar nature; and
>
> 3. adequately specified, or specified and billed, as appropriate prior to the invitation of tenders.

In view of the detailed provisions of this form and possible use on 'small' or 'minor' works, the Appendix to the JCT's Practice Note 20 is reproduced later comparing the Standard Form, Minor Works Form and the Intermediate Form.

.16

The Building Employers Confederation 'Guarantee Scheme' was also introduced in 1984 and a description with copies of its Form of Agreement and Conditions of Contract are included at the end of this Section. The Federation of Master Builders "Warranty Scheme" was introduced in 1981 and a description is also included at the end of this chapter.

4.17 The Finance (No 2) Act 1975

THE FINANCE (No 2) ACT 1975 – Statutory Deduction Scheme (Clause 31 of the "Standard" Form, Part C and Clause 5.3 of the "Minor Works" Form). In regard to this Scheme, the Board of Inland Revenue states that, within the meaning of Extract IR 14/15 (1976) the following are NOT Contractors:

1. a private householder having work done on his own premises (eg his home redecorated, repaired or extended).

2. a business, not normally involved in "construction operations" which does not have a separate department such as is mentioned in paragraph 16(a) and is simply having some work done for itself (eg a business having alterations or additions made to its premises.)

(Clause 16(a) re: the term "Contractor" includes:

> "some business and public bodies normally known in the industry as clients. This includes not only all local authorities but also those businesses and public bodies which maintain their own permanent building department, irrespective of its size, whose activities amount, in whole or part, to "construction operations" (see Appendices A and B of Extract). For Example, some large manufacturing, retailing and service concerns, and nationalised industries, engaged in "construction operations" on their own premises are "contractors".")

NOTE: With reference to Appendices A and B of Extract: Appendix A describes the construction operations which are included and excluded; Appendix B provides a list for guidance on which types of construction work are within the scheme and which are not.

4.18 OTHER FORMS OF CONTRACT that may be considered suitable, produced by certain professional organisations, include the:

A.C.A. Form – obtainable from
The Association of Consultant Architects,
7, Park Street,
Bristol BS1 5NF. Tel. No: 0272–293379.

F.A.S. Minor Works Form – obtainable from
The Faculty of Architects & Surveyors,
15, St. Mary Street,
Chippenham, Wilts.
Tel. No: 0249–444505.

4.19 Contents of JCT Contract Forms – Tabulated

STANDARD Clause Headings	MINOR WORKS Clauses	GRANT WORKS WITH Archt/SO Clauses	GRANT WORKS without Archt/SO Clauses	INTERMEDIATE FORM for Works of Simple Content
1980 Edition	Jan 1980 Issue	Dec 1975 Edition	Dec 1975 Edition	1984 Edition
1. Interpretation, Definitions, Etc.				The main differences in the Contract Conditions between the Standard Form, Minor Works Form and the Intermediate Form are set down following the Tabulations (paragraph 4.21 page 220)
2. Contractor's Obligations	1.1 Contractor's Obligation	1. Contractor's Obligations	1. Contractor's Obligations	
3. Contract Sum – additions or deductions – adjustments – Interim Certificates				
4. Archt/SO* Instructions	1.2 Archt/SO's Duties	2(i) Archt/SO Instructions		
5. Contract Documents – Other Documents – Issue of Certificates				
6. Statutory Obligations, Notices, Fees and Charges	5.1 Statutory Obligations, Notices, Fees and Charges	3. Statutory Obligations, (Notices) Fees and Charges		
7. Levels and Setting Out of Works				
8. Materials, Goods and Workmanship to conform to Description, Testing and Inspection				
9. Royalties and Patent Rights				
10. Foreman-in-Charge	3.3 Contractor's Representative	4. Contractor's Representative		
11. Access for Archt/ SO* to the works	3.4 Exclusion from Works			
12. Clerk of Works	3.5 Archt's SO's Instructions			
13. Variations and Provisional Sums	3.6 Variations 3.7 Provisional Sums	2(ii) Variations (iii) Provisional and Prime Cost Sums (iv) Addendum Price	7. Variations or Extra Work	
14. Contract Sum				
15. VAT – Supplemental Provisions	5.2 VAT	14. VAT	13. VAT	
16. Materials and Goods Unfixed or Offsite				
17. Practical Completion and Defects Liability	2.4 Completion Date 2.5 Defects Liability	9. Practical Completion – Defects Liability	4. Completion – Defects Liability	
18. Partial Possession by Employer				

STANDARD Clause Headings	MINOR WORKS Clauses	GRANT WORKS WITH Archt/SO Clauses	GRANT WORKS without Archt/SO Clauses	INTERMEDIATE FORM for Works of Simple Content
1980 Edition	Jan 1980 Issue	Dec 1975 Edition	Dec 1975 Edition	1984 Edition
19. Assignment and Subcontracts	3.1 Assignment			
19.A. Fair Wages – LA EDITION N/A	5.4 Fair Wages N/A			
20. Injury to Persons and Property and Employer's Indemnity	6.1 Injury to or death of persons 6.2 Damage to Property	7(i) Injury to or death of persons (ii) Damage to Property		
21. Insurance against injury to Persons and Property	6.4 Insurance – Evidence of Insurance	7(iii) Insurances – Persons and Property		
22. Insurance of the Property against Clause 22 perils	6.3.A. Insurance of the Works – Fire, etc – New Works 6.3.B. Insurance of the Works – Fire, etc – Existing	8. Insurance of the Works and Existing Structures – Fire etc.	8. Insurance of Works and Existing Structures against Fire, etc.	
23. Date of Possession, Completion and Postponement 24. Damages for Non-Completion 25. Extension of Time	2.1 Commencement and completion 2.2 Extension of Contract Period 2.3 Damages for Non-Completion 2.4 Completion Date 2.5 Defects Liability	6. Commencement, Progress and (Extension of Time)	2. Access (for commencement and Erection). 9. Extension of Period for Completion	
26. Loss and Expense Caused by Matters Materially Affecting Regular Progress of Works				
27. Determination by Employer	7.1 Determination by Employer	11. Determination by Employer	11. Determination by Employer	
(27.3 Corruption LA EDITION)	5.5 Prevention of Corruption			
28. Determination by Contractor	7.2 Determination by Contractor	12. Determination by Contractor	10. Determination by the Contractor	
29. Works by Employer or Persons employed or engaged by Employer				
30. Certificates and Payments	4.1 Correction of Inconsistencies 4.2 Progress Payments and Retention 4.3 Penultimate Certificate 4.4 Final Certificate 4.5 Contribution, Levy and Tax Changes 4.6 Fixed Price	10(i) Payments – Interim Payments (ii) Payments – Penultimate Certificate (iii) Payments – Final Certificate (iv) Payment of Grant to Contractor	3. Interim Payment 5. Final Payment 6. Payment of Grant to Contractor	

**SO limited to LA Editions. SO in recent revisions is termed "Contract Administration". N/A – no longer applicable.*

STANDARD Clause Headings 1980 Edition	MINOR WORKS Clauses Jan 1980 Issue	GRANT WORKS with Archt/SO Clauses Dec 1975 Edition	GRANT WORKS without Archt/SO Clauses Dec 1975 Edition
31. Finance (No 2) Act 1975 Statutory Tax Deduction Scheme. (Amdt issued November 1976)	5.3 (Finance (No 2) Act 1975) Statutory Tax Deduction Scheme (Amdt issued November 1976)	15.Finance (No 2) Act 1975 Statutory Tax Deduction Scheme (see Amdt RG3/1977 issued October 1977)	14. Finance (No 2) Act 1975 Statutory Tax Deduction Scheme (see Amdt XG3/1977 issued February 1977)
32. Outbreaks of Hostilities			
33. War Damage			
34. Antiquitiea			
Arbitration (see 'Articles' to the Standard Form)	Arbitration (see 'Articles' to the Agreement)	13. Arbitration	12. Arbitration
35. Nominated Sub-Contractors	3.2 Sub-Contracting	5. Sub-Contracting	
36. Nominated Suppliers			
37. Fluctuations			
38. Contribution, Levy and Tax Fluctuations			
39. Labour and Material Cost and Tax Fluctuations			
Supplemental Provisions (the VAT agreement)	8.1 Supplementary Memorandum (Part A – Tax, Etc, Charges: Part B – VAT; Part C – Statutory Tax Deduction)		

4.20 JCT Guidance notes on "Renovation Grant" forms

The following Notes for Guidance were published by the Joint Contracts Tribunal when the GRANT forms were first issued (following the Housing Act 1969):

Notes on the Agreement for Improvement Grant Works where an Architect/Supervising Officer is appointed and an Improvement Grant is to be made under the Housing Act, 1969. The Form is not for use in Scotland as advice has been received from the local authority organisations in Scotland that the existing procedure is adequate to cover work done under the Act in that country.

The note at the head of page 1 explains the kind of work for which the Form is to be used and that it is not appropriate where Bills of Quantities have been prepared or where a schedule of rates is required for valuing variations; for such work the Standard Forms (with or without Quantities) issued by the Tribunal should be used.

The main amendments to the existing Form of Agreement for Minor Building Works are as follows: the recitals make clear that the Form is only for use where an improvement grant is to be made. Clause 1 places an obligation on the architect/supervising officer to ensure that the works are so carried out as to secure the approval of the authority making the grant. Clause 2 deals with additions and omissions and contains a proviso that the architect/supervising officer has to advise the contractor if changes in the works will qualify for grant; the contractor is in these circumstances to be paid the reasonable costs of producing additional estimates for the purpose of getting additional grant unless the application is approved in which case the Contractor's costs are merged in the price payable for the additional works. Clause 2 also obliges the parties, where changes qualify for grant, to agree an inclusive price so that this price can be submitted to the authority for approval as the 'approved expenses' of the additional work. Clause 6 deals with commencement, progress and completion and provides for work to be carried out on every day except Sundays and for the parties to specify the hours when the premises will be made available for work. Clause 8 deals with insurance of the existing structures and the works against fire and certain other risks and contains a provision for the employer to produce evidence of adequate insurance having been taken out and maintained. Clause 10, dealing with payments, provides a new sub-clause under which (unless it is deleted by agreement) the employer authorises the authority to pay the grant, including any instalments, direct to the contractor and to enable this to be done copies of certificates of the architect/supervising officer have to be sent to the authority, the Appendix contains a Form of Authority which the employer has to sign to enable this direct payment to be made. Clauses 11 and 12 contain provisions for determination of the contractor's employment by the contractor and employer respectively. Clause 13 contains the arbitration agreement.

Notes on the Agreement for Building Works for which Improvement Grant (Housing Act 1969) is to be made where no Architect or Supervising Officer is appointed by the Employer. The Form is not for use in Scotland as advice has been received from the local authority organisations in Scotland that the existing procedure is adequate to cover work done under the Act in that country.

The recitals make plain that the contract for building works (as described or referred to in drawings and/or specification) is on the assumption that a grant has been approved and the amount of the 'approved expenses' and the amount of grant have to be inserted by the parties.

The agreement then provides for the works to be carried out within a period to be specified and inserted by the parties in the blank space provided and states the price to be paid; footnotes make clear that the price and the 'approved expenses' for grant purposes may not always be the same figure. The agreement refers to the conditions subject to which the works are to be carried out and which are set out following the signature of the parties.

Clause 1 sets out the contractor's obligations, including one to provide any necessary documents to get the authority's approval of the works. Clause 2 provides for the parties to insert the date of commencement and the actual period for completion will then be the period, calculated from the date of commencement, started in the agreement. Clause 3 provides for interim payments if the period for completion inserted in the agreement exceeds eight weeks and obliges the employer to make any necessary applications to the authority to enable grant to be paid. Clause 4 provides for a defects liability period of three months (unless another period is inserted in the conditions). Clause 5 provides for payment of 95 per cent of the total contract sum due at completion and for the balance to be paid when the defects have been rectified; it also obliges the employer to make any necessary application for release of the grant. Clause 6, unless it is deleted by agreement, provides for the employer to authorise the authority to pay the grant direct to the contractor and the Appendix contains an appropriate form of authority for this purpose. Clause 7 deals with additional works which the contractor has an option to carry out and for which further agreed terms, including price, will be necessary; if, however, the additional work required is, in the employer's view, eligible for grant then the contractor is to deal with any necessary documentation to enable an application for such grant to be made and this will be at the employer's expense unless the additional grant application is accepted by the authority. Clause 8 deals with insurance: the works and existing structures are at the employer's risk for fire, etc and the employer is obliged to insure in default of which the contractor may himself take out insurance at the employer's cost; the contractor is obliged to indemnify the employer, and take out insurance, against the risk of injury or damage to persons or property caused by the negligence of the contractor (or any person for whom he is responsible). Clause 9 provides for the completion period to be extended for causes outside the control of the contractor. Clauses 10 and 11 contain provisions for determination of the contractor's employment by the contractor and employer respectively subject to a proviso that such determination must not be made unreasonably or vexatiously. Clause 12 provides for arbitration and appointment of an arbitrator by the President of the Institute of Arbitrators.

4.21 Appendix to Practice Note 20 – Revised July 1984

Deciding on the appropriate form of JCT Main Contract

Reproduced by kind permission of RIBA Publications Ltd

Main differences in the Contract Conditions – Standard Form, Intermediate Form and Minor Works Form

Note: On the differences arising where bills of quantities are used or not used see paragraphs 15, 16 and 17 of the Practice Note.

(a) Date of Possession

Standard Form: provisions requiring possession of the site to be given to the Contractor on a date stated in the Appendix.

Intermediate Form: provisions similar to those in the Standard Form; in addition there is a provision enabling the Employer to defer possession for up to 6 weeks subject to the Architect making an extension of time if appropriate and a payment by the Employer to the Contractor for loss and expense, if any, incurred as a result.

Minor Works: provision that the Works "may be commenced" on a date to be inserted.

(b) Extension of Time

Standard Form: detailed provisions for notification by the Contractor to the Architect for matters affecting progress; the Architect to give decisions on fixing a new Completion Date in which he can take account of any variation requiring an omission ordered since a previous decision on fixing a new Completion Date; the grounds on which an extension of time can be given (the "Relevant Events") are set out in detail.

Intermediate Form: the events giving rise to an extension of time are similarly listed and are the same as those in the Standard Form except that there is no provision relating to the exercise of statutory powers by the UK Government or for delay on the part of nominated sub-contractors; and those events relating to the non-availability of labour or materials are optional. Additionally, there is an event relating to deferment of possession and there is specific provision regarding events occurring after the date for completion is past. The provisions for notification by the Contractor to the Architect for matters affecting progress are less detailed.

Minor Works: short provisions under which the Architect is notified by the Contractor if it becomes apparent that the Works will not be completed by the date for completion where this is due to "reasons beyond the control of the Contractor"; no notification by the Contractor needed if for any other reason; no definition of what is considered to be "beyond the control of the Contractor".

(c) Liquidated Damages

Standard Form: if the Architect certifies that the Works are not completed by the Completion Date the Contractor shall, if the Employer so requires, pay or allow to the Employer the liquidated damages at the rate stated in the Appendix; provision for adjustment if the Architect fixes a new Completion Date in his final review of extensions of time after Practical Completion.

Intermediate Form: similarly, if the Architect certifies that the Works are not completed on time the Contractor shall, if the Employer so requires, pay or allow to the Employer, liquidated damages at the rate stated in the Appendix; and there is also provision for adjustment should the Architect subsequently make an extension of time.

Minor Works: no certificate of the Architect as to whether the Works have been completed by the Completion Date. No provision for allowing the amount of the damages to the Employer (i.e. by deducting from monies otherwise due to the Contractor).

(d) Payment and Retention

Standard Form: detailed provisions on the compilation of Interim Certificates, the final account and the Final Certificate, on the treatment and trust status of Retention and on the payment of nominated sub-contractors; the legal effect of the Final Certificate is fully set out.

Intermediate Form: shorter but detailed provisions for the certification by the Architect of monthly interim and final payments, and on the effect of the final payment certificate; 95% of the value of work executed and materials on site is included in interim certificates ($97\frac{1}{2}$% at practical completion). The percentage not included is treated as trust money only where the Employer is not a local authority.

Minor Works: provision for progress payments where so requested by the Contractor in respect of the value of the Works properly executed and for goods and materials properly on site but without any further details on the compilation of those progress payments; retention not treated as trust money.

(e) Variations and Work in respect of Provisional Sums

Standard Form: detailed provisions, which where appropriate relate the valuation of variations to the rates and prices in the Contract Bills or the Schedule of rates.

Intermediate Form: broadly similar to those of the Standard Form.

Minor Works: valuation by the Architect "on a fair and reasonable basis using where relevant prices in the priced specification/schedules/schedules of rates"; no provisions as to when such prices are "relevant".

(f) Nomination of Suppliers

Standard Form: detailed provisions which are the same as those in the With Quantities Form.

Intermediate Form: no provisions.

Without Quantities: no provisions.

(g) Selection of Sub-Contractors by the Employer

Standard Form: detailed provisions enabling the nomination of Sub-Contractors by the Employer including the use of prescribed JCT Forms. For the Basic Method of nomination, these involve a Form of Tender (NSC/1), a Form of Employer/Sub-Contractor Agreement (NSC/2), a Form of Nomination Instruction (NSC/3) and the Form of Nominated Sub-Contract (NSC/4). For the Alternative Method the use of the Form of Sub-Contract NSC/4a, which includes an Appendix for inserting the information that would otherwise appear in the tender NSC/1 under the Basic Method, is obligatory. The use of the form of Employer/Sub-Contractor Agreement NSC/2a is also obligatory unless the tender has been invited on the basis that this Agreement is not required. A Form of Tender NSC/1a and a Form of Nomination Instruction NSC/3a are available but not obligatory.

The provisions dealing with the appointment by the Contractor of his own domestic sub-contractors also include the power of the Employer to require domestic work priced by the Contractor to be carried out by a sub-contractor chosen by the Contractor from a list of not less than three (which may be augmented by either the Contractor or the Employer in certain circumstances) provided by the Employer.

Intermediate Form: there is no provision for nominated sub-contractors; instead provision is made for work which is to be priced by the Contractor to be carried out by a person named in the Specification/Schedules of Work/Contract Bills, or in an instruction for the expenditure of a provisional sum, as a sub-contractor using the prescribed JCT Form of Tender and Agreement NAM/T. The consequences of default of the named person and the provisions for payment relating to the sub-contract works differ from those relating to nominated sub-contractors under the Standard Form.

There is no provision for a Contractor to choose a sub-contractor from a list of three provided by the Employer.

Minor Works: there are no provisions for either the nomination or naming of a sub-contractor nor for the selection by the Contractor of a sub-contractor from a list of three provided by the Employer. (Note: while it is possible for the Employer to seek to control the selection of sub-contractors for specialist work either by naming a firm or company in the tender documents or in instructions on the expenditure of a provisional sum, there are no provisions in the Form which deal with the consequences of what is in effect the nomination of a sub-contractor; nor is there any standard form of sub-contract which would be applicable to such selected sub-contractors.)

(h) Fluctuations

Standard Form: alternative provisions for contributions, levy and tax fluctuations or labour and materials cost and tax fluctuations. The further alternative of formula adjustment is available where there are Contract Bills.

Intermediate Form: the same provisions as for the Standard Form for contributions, levy and tax fluctuations but not for labour and materials costs and tax fluctuations; and the alternative, where there are Contract Bills, for the same formula adjustment as for the Standard Form.

Minor Works: Labour and materials and plant fluctuations expressly excluded from all payments including those for variations and provisional sum work; but the same provision for contributions, levy and tax fluctuations as in the Standard Form is included.

(i) Partial Possession

Standard Form: detailed provisions on possession by the Employer during the carrying out of the Works where this has been agreed by the parties; where sectional completion required from the outside of the contract a supplement of amendments to the Form is available.

Intermediate Form: no provisions, but where required the provisions for Partial Possession printed in Practice Note IN/1 may be used.

Minor Works: no provisions.

(j) Loss and/or Expense

Standard Form: detailed provisions, setting out the various matters which delay progress and cause the Contractor direct loss and/or expense; and the way in which such direct loss and/or expense is to be ascertained and paid; the matters relate to acts or defaults of the Employer or his Architect.

Intermediate Form: provisions similar to those in the Standard Form, with the addition of provision for reimbursement of loss and expense arising from deferment of the Date of Possession by the Employer.

Minor Works: no provisions and the parties are left to their common law rights with no contractual procedures for ascertainment and payment for damages due to acts or defaults of the Employer or his Architect which delay progress and cause the Contractor loss or expense for which he is entitled to damages.

(k) Arbitration

Standard Form: provision is made for disputes to be settled by arbitration, but, with certain exceptions, the reference to arbitration may not be opened while the Works are still in progress without the consent of both parties; the arbitrator's powers include revising certificates or decisions of the Architect.

Intermediate Form: similar provisions as for the Standard Form, but without the restraint on opening references to arbitration while the Works are still in progress; in addition, an optional provision under which questions of law are referable to the High Court.

Minor Works: similar provision for appointing an arbitrator as in the Intermediate Form, but it is not expressly stated that the arbitrator's powers include revising certificates or decisions of the Architect.

Other differences include:

(l) Testing

The Standard and Intermediate Forms contain similar provisions enabling the Architect to issue instructions for inspection or tests of work or materials, but the Minor Works Form has no such provision.

In addition, where a failure of work or materials to be in accordance with the Contract is discovered during the carrying out of the Works, the Intermediate Form makes provision for the opening up for inspection or testing of similar work or materials by the Contractor at no cost to the Employer, if such opening up or testing is reasonable in all the circumstances.

(m) Insurance

The provisions of the Standard Form relating to insurance are followed in the Intermediate Form. The provisions of Minor Works are less detailed and in particular do not include the provision for the Contractor to insure the Employer's liability to others for damage to property other than the Works.

(n) Determination

Both the Standard Form and the Intermediate Form contain detailed and similar (but not identical) provisions relating to determination of the Contractor's employment under the contract. Under the Intermediate Form, the Contractor is not entitled to payment for direct loss and/or damage resulting from determination consequent upon: force majeure, loss or damage to the Works occasioned by the specified perils, or civil commotion. The Minor Works provisions are less detailed.

(o) Payment for Unfixed Materials on Site

The Standard and Intermediate Forms contain the same provisions requiring the Contractor to include in any sub-contract certain conditions relating to the passing of property in unfixed materials and goods on site when paid for by the Employer. There is no such provision in the Minor Works Form.

4.22 The BEC Guarantee Scheme

The BEC Building Trust Ltd has provided the following information about its "Guarantee Scheme" that is reproduced below with its kind permission:

The Guarantee Scheme was initiated by the BEC (Building Employers Confederation) in October 1984 to provide protection and benefits over and above those normally available under contracts for work of a general building nature. The Scheme is administered, on behalf of the BEC, by the BEC Building Trust Ltd (the Trust) and is *insurance* backed.

The builder, or specialist, must be a Member of the BEC when the work is registered for Scheme cover with the Trust.

Initially, the Guarantee Scheme only covered contracts where the BEC Member had been employed directly by the Customer and a professional had not been engaged to act, on the Customers behalf, whilst the work was being carried out – the **Standard Scheme**. In this case the Customer was not, however, precluded from using professional services for the preparation of plans/specifications or obtaining planning permissions, etc. A variant to the Standard Scheme was introduced in October 1987, designed specifically to provide a Guarantee Scheme cover where an Architect or a Contract Administrator had been appointed to act on behalf of the Client – the **Supervised Scheme**.

The BEC Guarantee Scheme covers jobs, currently in the range £500–£100 000 under the Standard Scheme and £500–£125 000 under the Supervised Scheme. It applies to all normal building work, both private and commercial, including: conversions, extensions, kitchen work, painting and decorating, central heating and electrical installations, etc. It does not cover contracts predominantly for repairs to existing roofs; landscape gardening; swimming pools and work connected therewith; or solar heating.

In brief terms, the Scheme provides a safeguard if things go wrong, including:

> A formal written contract: the special Agreement for Scheme Work where Customers act on their own behalf; the JCT Agreement for Minor Building Works with Supplementary Memorandum E if a professional is acting for the Client.
>
> A guarantee that the work will be completed in accordance with the contract. If the BEC Member becomes insolvent, the Trust will ensure that another Confederation Member finishes the job. Any

additional cost up to £10 000 (Standard Scheme) or £12 500 (Supervised Scheme), including VAT and Professional Fees, will be met by the Scheme.

The means whereby problems and disputes, under the Standard Scheme, can be resolved quickly and informally by an independent professionally qualified conciliator, at no cost to either the Customer or the Builder (except a refundable deposit), followed by an arbitration service if the conciliation is not successful. The JCT contract provides for arbitration under the Supervised Scheme.

That for six months after the job has been finished, any work which has not been done properly will be put right by the BEC Member.

An additional assurance that structural defects in the foundations or load-bearing parts of any roof, floor or wall, which arise from the work within the following two years, and for which the BEC Member is responsible under the Scheme, will be rectified. If the Member has become insolvent the Scheme will meet costs up to £10 000 under the Standard Scheme and £12 500 under the Supervised Scheme, including VAT and Professional Fees.

Full insurance cover for damage to the work while it is being carried out. Limitations, as with any insurance, are explained in the Scheme documentation.

For the Guarantee cover to apply, a registration fee of 1% of the total contract price, excluding VAT, is payable (minimum fee £20.00).

Details of the Scheme together with a list of BEC Members who carry out work in the enquirer's locality can be obtained from the:

BEC Building Trust Ltd,
18 Mansfield Street
London
W1M 9FG
Tel: 071 580 6306

The following copies of the 'Agreement for Scheme Work' and 'Conditions of Contract', under the Standard Scheme, are reproduced with the kind permission of the BEC Building Trust Ltd. Copies of the JCT Agreement for Minor Building Works, for use under the Supervised Scheme, can be obtained from: BEC Publications, 2309 Coventry Road, Sheldon, Birmingham B26 3PL. Tel: 021 742 0824.

4.23

GS/2
Agreement for Scheme Work

BEC Guarantee Scheme

BEC GUARANTEE SCHEME

AGREEMENT
Between ______________________
______________ ('the Customer')
of ___________________________

Dated______________ 19 ___
and ___________________________
______________ ('the Member')
of ___________________________

RECITALS

First The Customer requires the following Scheme Work to be carried out:
[**a**] ______________________
______________ ('the Work')
at ___________________________
(give site address if different from Customer's address):

Second The **Member** has provided Drawings numbered____and/or a Specification dated______ (copies of which have been passed to the Customer) showing and describing the Work [**b**]:

Third The **Customer** has provided Drawings numbered____and/or a Specification dated______ (copies of which have been passed to the Member) showing and describing the Work [**b**]:

Fourth All necessary statutory permissions, consents and approvals have been obtained as follows: [**c**]

1 Planning permission No ______granted under the Town & Country Planning Act, 1971, Part III:
.1 full planning permission including approved detailed drawings, or (Yes/No/ Not applicable)
or
.2 outline planning permission together with subsequent approval of detailed matters pursuant to this outline (Yes/No/Not applicable)

2 Listed building consent (where applicable) granted under the Town & Country Planning Act, 1971, Part IV (Yes/No/Not applicable)

3 Building and Fire Regulation approval Nos____
____
____
granted under the Public Health Acts 1936–1961 or other relevant legislation (Yes/No/Not applicable)

4 Other relevant documents
____
____
and both Customer and Member have the copies of such permissions, consents, approvals or other documents.

Fifth The Customer has received in Booklet GS 9 a copy of:
– the Conditions of Contract

GS/2

Agreement for Scheme Work

BEC Guarantee Scheme

('the Conditions') referred to in clause 2 of this Agreement:

- the 'Scheme Protection and Guarantee' referred to in Condition 9 of the Conditions:
- the Scheme Rules.
- the terms of the insurance referred to in Condition 13.2 of the Conditions.

Sixth The Customer and Member have completed the Guarantee Registration Application (GS 5) requesting BEC Building Trust Ltd to apply the BEC Guarantee Scheme to this Agreement.

TERMS OF AGREEMENT

1. This Agreement shall come into effect only and at the date when BEC Building Trust Ltd. has confirmed in the Registration Certificate (GS/7) that the BEC Guarantee Scheme shall apply.

2. Subject to the Conditions the Member shall carry out and complete the Work in accordance with the Drawings and/or Specification referred to in the Second and/or Third recital.

3. The Customer shall pay to the Member the Contract

   Price of £______(exclusive of VAT) or such other amount as shall become payable in accordance with the Conditions (the 'Final Contract Price').

4. Value Added Tax ('VAT') shall be charged by the Member on the value of any part of the Work for the Supply of which the Member is statutorily liable to account for VAT and the Customer shall pay such VAT to the Member as provided in the Conditions. [**d**]

5. The Conditions shall have effect subject to the following particulars:

	Condition
.1 date of access for Member to the site of the Work______	2.1
.2 date for completion of the Work______	2.1
.3 facilities, if any, to be provided by Customer – Health & Safety at Work Act 1974	12.1
.4 the Customer shall provide for use in connection with the Work the following services	12.2

Signed by:

____________________ __________

For and on behalf of the Customer — Capacity

____________________ __________

For and on behalf of the Customer — Capacity

____

GS/2

Agreement for Scheme Work

BEC Guarantee Scheme

[**a**] Insert brief description of the work required.

[**b**] If both the Customer and Member have provided Drawings and or a Specification, retain both recitals and identify the documents provided by each. Otherwise delete Second or Third recital as appropriate.

[**c**] Complete as appropriate

[**d**] Not all building work is subject to standard rate VAT. The application of VAT to building work is set out in the relevant H.M. Customs & Excise Notices.

4.24

GS/3 **Conditions of Contract**

Condition

OBLIGATIONS OF MEMBER

Carrying out and completion of the Work

1 The Member shall execute the Work with reasonable skill and care and in accordance with the Drawings and/or Specification referred to in the Second and/or Third recital of the Agreement.

Access – hours when site available – date for completion

2.1 The Customer shall, from the date of access stated in the Agreement, grant to the Member adequate and unimpeded access to the site of the Work on every day, excluding Sunday, for the purpose of carrying out the Work and shall also provide adequate working and storage space. The Member may thereupon commence the Work, and shall proceed regularly and diligently to complete the Work on or before the date for completion stated in the Agreement or fixed under Condition 2.2.

Extension of time for completion of the Work

2.2 If for any reason beyond the control of the Member, the Work will not be completed by the date for completion, the Member shall so notify the Customer and fix a revised date in substitution for the date for completion stated in the Agreement.

Design of the Work – liability where Member undertakes design

3 Where and to the extent only that the Member has undertaken the design of the Work (including any variation thereof under Condition 11) the Member shall be under the same liability in respect of any defect or insufficiency in such design as would an appropriate professionally qualified practioner in the like circumstances.

Specification or supply of materials by Customer

4.1 Where the Customer specifies or supplies materials or goods which, in the opinion of the Member, are not suitable for the purpose for which they are required, the Member shall accordingly notify the Customer in writing setting out the reasons why such materials or goods are, in his opinion, unsuitable.

4.2 If, on receipt of the notification referred to in Condition 4.1, the Customer nevertheless requires that the said materials or goods shall be used in the carrying out of the Work or does not otherwise rely on the skill and

GS/3 **Conditions of Contract**

4.2 *continued*

judgment of the Member, no responsibility for their suitability is accepted by the Member; nor shall the Member be liable for any loss or damage caused during or by virtue of their incorporation or fixing in the work, save where such loss or damage is caused by the negligence (as defined in the Unfair Contract Terms Act 1977) of the Member.

Sub-Contracting – notification

5 The Member may sub-contract any part of the Work and he shall notify the Customer of the name of the sub-contractor or sub-contractors.

PAYMENT

Interim Applications for Payment

6.1 The Member, at intervals of not less than 2 weeks (calculated from the date of access stated in the Agreement), may make an Interim Application for Payment to the Customer in accordance with the provisions of Condition 6.2.

Computation of Interim Applications for Payment

6.2 An Interim Application for Payment shall state:

.1 the total value of work properly executed up to the date of the Application for Payment including the value of any materials or goods delivered to the site of the Work for incorporation in the Work;

less

.2 the amounts previously paid in respect of earlier Interim Applications for Payment

together with

.3 any amount due in accordance with the Agreement in respect of VAT on the value of work included in the Interim Application for Payment.

Payment by Customer of Interim Application for Payment

6.3 The Customer shall pay to the Member within 14 days of the date of issue of each Interim Application for Payment the amount shown as due on that Application.

6.4 The Member shall not remove from the site of the Work any materials or goods for which the Customer has paid under Condition 6 without the permission of the Customer.

GS/3 Conditions of Contract

PRACTICAL COMPLETION AND FINAL PAYMENT

Practical completion notification – Final Application for Payment – Final Statement of VAT

7.1 When the Work has reached practical completion the Member shall:

.1 so notify the Customer in writing:

.2 send to the Customer, not later than 3 months after the date of practical completion a Final Application for Payment setting out
– the Contract Price;
– any adjustments to the Contract Price to arrive at the Final Contract Price;
– the Final Contract Price
and stating the amount or balance due to the Member or to the Customer, as the case may be, after deducting (where applicable) from the Final Contract Price any amounts previously paid by the Customer.

.3 send to the Customer a Final Statement of the total VAT due in respect of the Work and showing the balance due to the Member or to the Customer as the case may be after deducting therefrom (where applicable) amounts on account of VAT previously paid by the Customer.

.4 calculate any reduction in fee payable by the Customer under Rule 4 of the Scheme Rules and give credit in the Final Contract Price to the Customer for such reduction.

.5 as agent for the Trust calculate any increase in fee payable by the Customer under Rule 4 of the Scheme Rules, add such increase to the Final Contract Price and pay that increase to the Trust as agent for the Customer.

.6 send to the Trust Confirmation Form (GS/8) confirming to the Trust the date of practical completion, the VAT-exclusive Final Contract Price and its payment by the Customer together with a statement of any addition or reduction in the fee payable by the Customer and a remittance to the Trust for any such addition.

Payment of Final Application

7.2 Subject to Condition 7.4 the Customer or Member shall, within 14 days of the issue of the Final Application for Payment, pay to the other the balance shown in that Application.

Final Payment of VAT

7.3 Within 14 days of the issue of the Final Statement of the total VAT due in respect of the Work the

GS/3 **Conditions of Contract**

7.3 *continued*

Customer or the Member shall pay to the other the balance shown in that Statement.

7.4 The Final Application for Payment shall, 14 days after its issue, be conclusive, save for any accidental inclusion or exclusion of any work, goods, materials or figure in any computation, or any arithmetical error in any computation, as to the balance stated therein, unless within that period the Customer gives notice in writing to the Member that he considers the balance (or any part thereof) to be incorrect. In such case the balance shall be conclusive to the extent agreed by the Customer and Member.

RECTIFICATION OF DEFECTS

Defects – making good by Member

8.1 Any defects which appear within the defects liability period (that is the period of 6 months from the date of practical completion as notified under Condition 7.1.1) which are due to materials and workmanship not being in accordance with the contract or to frost occurring before the date of practical completion shall be made good by the Member as soon as possible after written notification by the Customer to the Member of such defects.

Completion of making good defects – notification to the Customer

8.2 The Member shall notify in writing the Customer of the date when in the opinion of the Member the Member's obligations under Condition 8.1 have been discharged.

BEC GUARANTEE SCHEME

9 Subject to Rule 8 of the Scheme Rules the Customer and the Member agree to abide by and be bound by the 'Scheme Protection and Guarantee' (GS/4) operated by BEC Building Trust Ltd and be responsible to BEC Building Trust Ltd in respect of any matters covered by that Guarantee.

PROVISIONAL SUMS

Meaning of Provisional Sum

10.1 Where the Work involves work or costs which cannot be entirely foreseen, defined or detailed at the date of the Agreement a provisional sum or sums shall be included in the Specification for the Customer ('Customer's Provisional Sum') or by the Member ('Member's Provisional Sum).

Customer's Provisional Sums

10.2 The Customer shall notify the Member of his decision on the

GS/3 Conditions of Contract

10.4 *continued*

expenditure of any Customer's Provisional Sum within a reasonable time after the date of the Agreement so as not to delay the progress of the Work. The Member shall confirm in writing to the Customer his acceptance of that decision.

10.3 The Customer shall pay instead of any Customer's Provisional Sum the cost of the supply of any materials, goods or workmanship to which that Sum refers together with a fair and reasonable amount in respect of the overheads and profit of the Member.

Member's Provisional Sums

10.4 Where reasonably practicable the Member shall notify in writing the Customer of the amount to be paid by the Customer in respect of a Member's Provisional Sum before any expenditure is incurred in respect of the items to which such Sum refers. The Customer shall, subject to Condition 10.5, notify in writing the Member of his acceptance of the obligation to pay that amount and shall be entitled to any reasonable breakdown of that amount.

10.5 If on receipt of any notification under Condition 10.4 the Customer considers that he will not be able to pay the amount the Customer and Member shall agree any reduction in the work for which the Member's Provisional Sum was included and in the amount to be paid. Provided that if such reduction would adversely affect the satisfactory carrying out of the Work the Member may at his discretion either terminate the Agreement or agree on the reduction and any consequential limitation of his liability in respect of the satisfactory nature of the Work which may be fair and reasonable.

10.6 Where it was not reasonably practicable to notify the Customer under Condition 10.4 of the amount to be paid by the Customer in respect of a Member's Provisional Sum before any expenditure is incurred, the Member shall nevertheless keep the Customer informed of the amount of any expenditure incurred or being incurred; and shall provide the Customer with any reasonable breakdown of the amount of such expenditure. Subject to these obligations of the Member the Customer shall pay the amount properly expended in respect of a Member's Provisional Sum to which Condition 10.6 is applicable.

10.7 The payments referred to in Conditions 10.3, 10.4, 10.5 and

GS/3 Conditions of Contract

10.7 *continued*

10.6 shall be added to or deducted from the Contract Price and included in Applications for Payment.

VARIATIONS

11.1 .1 The Member shall if authorised in writing by the Customer vary the Work by the addition, omission or substitution of any work (a 'Variation') and no Variation shall vitiate this Agreement.

.2 The Member shall carry out any Variations to the Work which are necessary for compliance with any relevant stautory obligations.

11.2 Wherever practicable the price for any Variation shall be agreed before the Variation is carried out. Where such agreement is not reached the price shall be such amount as is fair and reasonable having regard to the nature of the Variation, the circumstances in which the Variation was authorised by the Customer and carried out by the Member, and its effect on remaining work; provided that the Customer shall not pay for any Variations within Condition 11.1.2 where the statutory obligation became law before the date of the estimate which resulted in the execution of the Agreement.

11.3 The price for Variations shall be added to or deducted from the Contract Price as the case may be and included in Interim Applications for Payment.

USE OF SERVICES

12.1 The Customer shall provide the Member with the facilities required under the Health & Safety at Work Act 1974 or as may otherwise have been agreed.

12.2 The Customer shall provide for use in connection with the Work the services set out in clause 5.4 of the Agreement.

INSURANCE OF THE WORK

Insurance of the Work against Fire etc

13.1 The BEC Building Trust Ltd. shall in the joint names of the Member and the Customer insure against loss or damage to all work executed and all unfixed materials and goods delivered to, placed on or adjacent to the Work and intended for incorporation therein.

13.2 The terms and duration of the Contract Works Insurance referred to in Condition 13.1 and

GS/3 **Conditions of Contract**

13.2 *continued*

the amount of indemnity given thereunder are set out in the Booklet GS/9 issued by BEC Building Trust Ltd. a copy of which has been given to the Customer.

13.3 The BEC Building Trust Ltd. shall produce such evidence as the Customer and/or the Member may reasonably require that the insurance referred to in Conditions 13.1 and 13.2 has been effected and is in force at all material times.

13.4 Subject to the provisions of Condition 13.1, the Member shall be responsible for and shall make good or restore all loss or damage to all work executed and all unfixed materials and goods delivered to, placed on or adjacent to the Work and intended for incorporation therein until the date of issue of the Final Application for Payment under Condition 7 except to the extent that such loss or damage is caused by the act or default of the Customer or any person for whom the Customer is responsible.

13.5 The BEC Building Trust Ltd shall be responsible for the payment out of any monies due under the insurance to which Condition 13 refers.

INJURY TO PERSONS – DAMAGE TO PROPERTY – INSURANCE

Injury to persons – Member's liability and indemnity

14.1 The Member shall be liable for, and shall indemnify the Customer against, any expense, liability, loss, claim or proceedings whatsoever arising under any statute or at common law in respect of personal injury to or the death of any person whomsoever arising out of or in the course of or caused by the carrying out of the Work except to the extent that the same is due to any act or neglect of the Customer or any person for whom the Customer is responsible.

Damage to property – Member's liability and indemnity

14.2 The Member shall, subject to Condition 13.1, be liable for, and shall indemnify the Customer against, any expenses, liability, loss, claim or proceedings in respect of any damage whatsoever to any property real or personal other than the Work insofar as such damage arises out of or in the course of or by reason of the carrying out of the Work, and to the extent that the same is due to any negligence, omission or default of the Member or of any person for whom the

GS/3 Conditions of Contract

14.2 *continued*

Member is responsible or of any sub-contractor employed by the Member or of any person for whom such sub-contractor is responsible.

Insurance

14.3 Without prejudice to the Member's liability to indemnify the Customer as stated, the Member shall effect, and shall cause any sub-contractor to effect, such insurances as are necessary to cover the liability of the Member or, as the case may be, of any sub-contractor, for the personal injury to or death of any person, or damage to property referred to in Conditions 14.1 and 14.2.

14.4 The insurances required by Condition 14.3 shall provide for cover in an amount not less than £500,000 for any one occurrence or series of occurrences arising out of one event except that in respect of liability for the death of or injury to any person under a contract of service or apprenticeship with the Member and arising out of and in the course of such person's employment by the Member, the amount insured shall comply with the Employer's Liability (Compulsory Insurance) Act 1969 and any statutory orders made thereunder or any amendment or re-enactment thereof.

14.5 The Member shall produce such evidence as the Customer may reasonably require that the insurances referred to in Conditions 14.3 and 14.4 have been effected and are in force at all material times.

CLEARANCE OF SITE

15 During the carrying out of the Work the Member shall keep the site reasonably clear of all rubbish and debris resulting from the execution of the Work, and on completion shall clear the site of all such rubbish and debris to the reasonable satisfaction of the Customer.

STATUTORY OBLIGATIONS

16.1 The Member shall, unless otherwise agreed between himself and the Customer, give all notices required to be given in connection with Building Regulations and with the supply of any necessary services and pay all fees thereby legally demandable by any relevant authority or company.

16.2 The Member shall not be liable to the Customer if the work does not comply with the statutory requirements where and to the

GS/3 Conditions of Contract

16.2 *continued*

extent that such non-compliance of the Work results from the Member having carried out work in accordance with any documents which have been provided by the Customer and referred to in the Third recital of the Agreement (GS 2).

BECOMING INSOLVENT – DETERMINATION

17.1 If the Member becomes insolvent, as defined in the Scheme Protection and Guarantee (GS/4) Section 9.1.3, the employment of the Member under the Agreement is forthwith automatically determined.

17.2 Upon such determination the Customer may, subject to the terms of the Scheme Protection and Guarantee, employ and pay other persons to carry out the Work and the Member shall forthwith remove from the Work and the site any plant, tools, equipment, goods and materials belonging to or hired by him.

17.3 Until the completion of the Work in accordance with the relevant provisions of the Scheme Protection and Guarantee the Customer shall not be bound to make any further payment to the Member. Upon such completion the Customer shall be paid by the Member the amount of any direct loss and/or expense caused to him by the determination less any amounts properly due under the Agreement to the Member and the resultant sum shall be a debt payable by the Member to the Customer or by the Customer to the Member as the case may be.

CUSTOMER'S STATUTORY RIGHTS

18 Nothing in the Agreement (GS/2) or in these Conditions shall be construed so as to reduce or in any way modify any rights to which the Customer may be or become entitled under any statutory provisions.

4.24

The FMB Warranty Scheme

This scheme was initiated by the Federation of Master Builders in 1981 to achieve better recognition of its members services and service to their clients through an insurance-backed scheme for works of improvement, extension and repairs; generally up to a value of £75 000 (inc. VAT).

The members are registered by the FMB on its National Register of Warranted Builders for which they have to be approved. This means they have been in business for a minimum of three years with continuous trading experience; are able to supply a selection of three recent satisfactorily completed jobs which are checked by the Registration Board; can give details of the firm's structure and financial turnover; hold current policies for Employers, Public and Product Liability Insurances; can provide satisfactory banker's references and are able to pay the insurance premium for the warranted work.

Their membership is recognised by their entitlement to display the special insignia that resembles a shield with a bold WR in its centre and "National Register of Warranted Builders" around the rim.

For jobs covered by the Warranty Scheme a premium of 1% is charged on the cost of a job (plus VAT).

Full particulars and forms can be obtained from;

The Registrar,
National Register of Warranted Builders,
Federation of Master Builders,
Gordon Fisher House,
33, John Street,
London WC1N 2BB

An explanation by the FMB is aptly captioned "choosing a builder should be as carefully thought over as choosing your next car or where to go for your next holiday" and is given below by kind permission of the Registrar of the Scheme.

How to go about it

Almost certainly it is going to be expensive, so it is best to make sure before you start that the person you choose is well qualified. Also, it is in your own interest to look for some form of guarantee so that, if for some reason the workmanship or materials used proved faulty, you would have the necessary backing to make sure it was put right without further cost to you.

In the first place, look for someone who is a member of a recognised builders' organisation, such as the Federation of Master Builders. Although in itself this does not give a guarantee, it does prove that he is a man of experience and that, in becoming a member, he has undertaken to observe the principles and rules laid down by the Federation.

Next, *for real peace of mind,* look for a Federation member who is also on the National Register of Warranted Builders. The Warranty Register is a scheme operated by the Federation of Master Builders in response to a demand from the public for greater safeguards in the face of growing numbers of "scare" stories about work carried out by "cowboy" builders which proved to be unsatisfactory – sometimes even dangerous – and was very expensive to have put right.

Protection through the Warranty Scheme

Whatever the job in question – plumbing, plasterwork, building, carpentry, or laying crazy paving even – it can be covered under the scheme up to a ceiling figure of £75 000 (inc. VAT), using a warranted builder. Provided that it has been negotiated with the builder before the work is started, the warranty will still apply even if you choose to have an architect or surveyor act as supervisor or co-ordinator on the job.

The warranty guarantees you:

- protection whilst the work is in progress, meaning that, should the builder cease trading through insolvency or death of the sole proprietor before the job has been fully and satisfactorily completed, any proven additional costs incurred through having to find another builder to complete the contract, will be paid by the Federation to a maximum of £10 000.
- that the completion work will be undertaken by another registered builder of the client's choice.
- the satisfaction of knowing that, for a period of two years after the work has been completed, defects through faulty materials or workmanship can be reported in writing to the Registrar and, if agreed, will be rectified at no expense to the client even if the builder has gone bankrupt in the meantime, or – if sole proprietor – has died.
- that acknowledged defects will be rectified as quickly as possible.
- that the warranted builder holds current policies for Employers, Public and Product liability insurances.
- in the event of a dispute between a client and a Warranted builder, the Registration Board provides a free conciliation and arbitration facility to reach a speedy and inexpensive decision. If the decision is reached in favour of the client, the Registrar instructs the builder to remedy the defects. Should he refuse, the Board will pay the reasonable costs of having the defects made good by another builder chosen by the client.

Legal rights

Clients who have warranty registered work done for them retain all their legal rights. The Warranty Scheme actually supplements these, particularly in relation to its provisions on insolvency or death of a sole proprietor.

Reasonable costs
The Warranty Scheme undertakes to meet all reasonable costs of rectification of a defect by another builder of the client's choice in respect of the following:

(a) where a builder refuses to honour his guarantee under the scheme
(b) death during the warranty period of the registered builder if he was sole proprietor.
(c) the cessation of trading by the builder due to bankruptcy, creditors voluntary liquidation or compulsory liquidation during the warranty period.

The original warranty period will continue to be observed in these cases.

Insurance backed
The Warranty Scheme, as operated by the Federation, is an insurance-backed scheme negotiated with Minet Insurance Brokers (UK) Limited. The only cost to the client for this strong guarantee and the additional peace of mind that comes with it, is just 1% of the contract price (£5 on a £500 job, minimum fee £5).